OSCAR:
The Ham Radio Satellites

Other TAB books by the author:

No. 859 *The Complete Handbook of Slow-Scan TV*

No. 1120

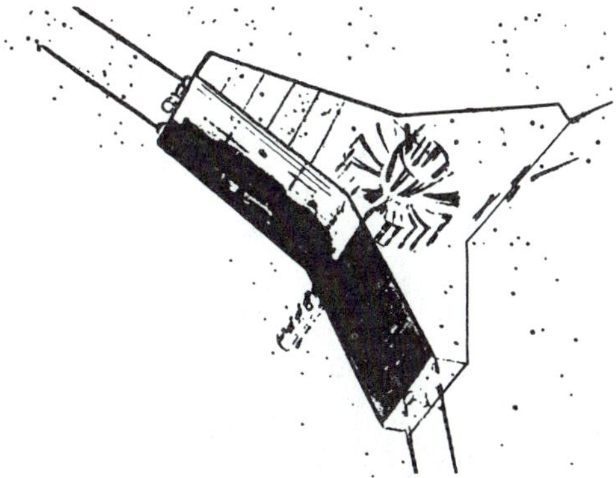

OSCAR: The Ham Radio Satellites

by Dave Ingram, K4TWJ

TAB BOOKS
BLUE RIDGE SUMMIT, PA. 17214

FIRST EDITION

FIRST PRINTING—DECEMBER 1979

Printed in the United States of America

Library of Congress Cataloging in Publication Data

Ingram, Dave.
OSCAR: the ham radio satellites.

Includes index.
1. Amateur radio stations. 2. OSCAR (Artificial satellite) I. Title.
TK9956.I514 621.3841'97 79-12614
ISBN 0-8306-9805-1
ISBN 0-8306-1120-7 pbk.

The cover photo is a conception by artist William R. Findley WA6TUF of OSCAR 8 in orbit.

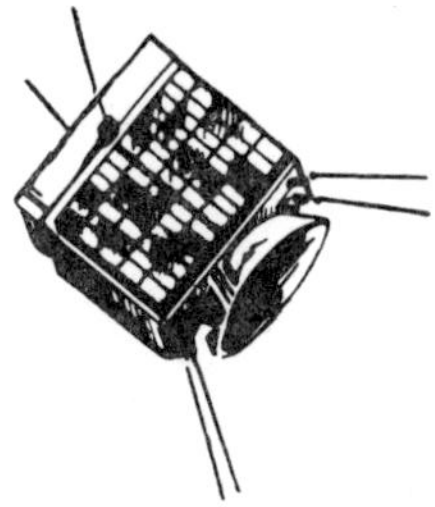

Preface

Space communication is one of the hottest new frontiers in amateur radio, and its popularity is rapidly increasing. Each day, amateurs around the world are experiencing the enjoyment of this new era and realizing the fascination in personal communication via OSCAR satellites. There are a number of unique aspects associated with satellite communication, and these features have given many amateurs a renewed enthusiasm for ham radio. If you're searching for more than the usual pleasures in amateur radio, I'm sure you, too, will find OSCAR activities an absolute blast.

Although amateur satellites have been flying since 1961, their use hasn't become widespread until recent times. This enthusiasm has obviously been escalated by our latest satellite, OSCAR 8, and the upcoming Phase III spacecrafts. The construction cost of amateur radio satellites has been borne entirely by the amateur fraternity. Fortunately, these satellite costs have been significantly less than their commercial counterparts. As an example, OSCAR 7 costs totalled approximately $64,000 whereas a similar unit built under industrial auspices would have cost $2 million. The American Radio Relay League (ARRL) has made a substantial financial contribution to AMSAT in support of OSCAR 8. Education materials for OSCAR users also are being widely distributed by the ARRL.

This book is intended as an introduction and guideline for amateurs investigating or joining the fun of satellite operations. It is purposely *enjoyment* oriented rather than technically oriented, so settle back in an easy chair with a Coke and start dreaming of your own satellite earth station.

Writing a book of this nature is not a simple task, and I would like to express sincere gratitude to the following individuals for their outstanding assistance and support in producing this OSCAR statellite book: The Radio Amateur Satellite Corporation (AMSAT), and its U.S. president, Dr. Perry Klien, W3PK; Dr. Norman Chalfin, K6PGX; Mike Smithwick, WA6TUF; Roy Welch, WØSL; and Fred Merry, W2GN. Thanks also to the following companies for their assistance with information supplied in Chapter 7 of this book: ARCOS, Cushcraft, Hamtronics Inc., Spectrum International and Yaesu Electronics.

Finally, my deepest appreciation and gratitude to my XYL, Sandy, WB4OEE, for typing this manuscript and standing by me while writing this book.

As with any state of the art book which deals with pioneering frontiers, timeliness is a problem. Operational parameters and schedules are always open for modification; thus, late issue periodicals should supplement information presented in this publication.

Dave Ingram, K4TWJ

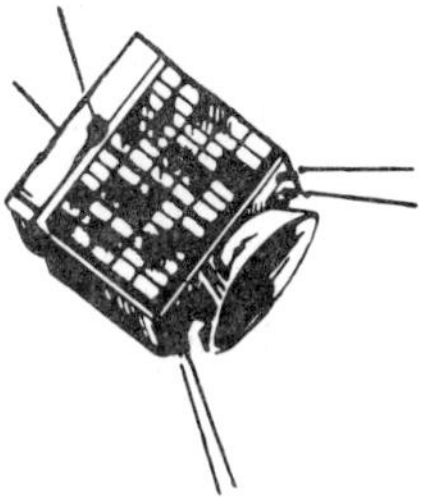

Contents

Chapter 1
Introduction to Satellite Communication

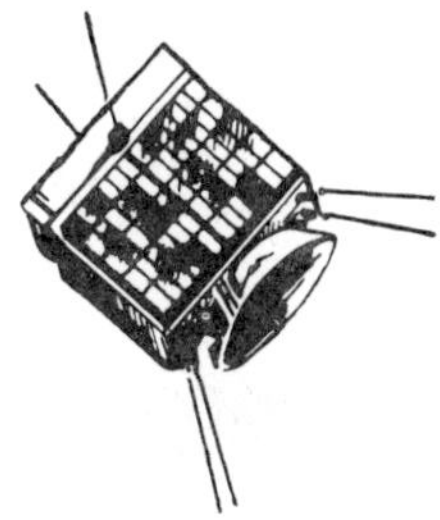

Communication via satellite is a frontier of unlimited opportunity and enjoyment that's open to amateurs around the world on a daily basis. This exciting new era of amateur radio is gaining momentum at a fantastic rate, and some revolutionary developments are due to evolve in the near future. A superb new amateur radio satellite recently was placed in orbit around our world (Fig. 1-1), and another highly sophisticated amateur satellite capable of direct worldwide communication on a multi-hour basis is scheduled for launch during early 1980. Now's the ideal time to hop on the bandwagon and join in the fun of this new aspect of amateur communications. This book is intended as a guideline toward that end, and it provides all the necessary details you need to become involved with today's space age communications. I trust you'll find the information useful, entertaining and enjoyable.

It has been said that our Golden Age of Communications has, like many aspects of the exciting yesteryears, passed in the annals of time. When we consider that situation from a different standpoint, however, we find the Golden Age is still very much alive in the special frontiers of advanced communication techniques. OSCAR (Orbital Satellite Carrying Amateur Radio) satellite communications are a perfect example of such boundless horizons. If you're looking for a refreshing change in amateur radio activities, this mode will surely provide such new life and return that excitement you felt during those first days as a ham radio operator.

The launch of U.S.S.R's Sputnik I in late 1957 sparked interest and enthusiasm among radio amateurs throughout the world. Soon

Fig. 1-1. The newest star performer in amateur radio's satellite program, OSCAR 8, sits on its mount baseplate awaiting final checkout at Vandenburg Air Force Base laboratories. Outer perimeter of satellite is covered with high quality photocells. White blocks at top center of craft is outlet for 10 meter antenna.

thereafter, a relatively small group of amateurs began building a space satellite of their own. This "weekend project," which was informally accomplished in basements and garages, became our first orbital amateur radio satellite in late 1961. The satellite program immediately acquired acceptance and soon became too large for a single, small group to handle. AMSAT, the Amateur Satellite Corporation, was thus established and large scale progress began. Today, AMSAT continues the design, development and launch aspects of amateur satellites, while the American Radio Relay League assists with many administration and public relation duties of the spacecrafts. Our satellite program is now moving forward, and its possibilities for the future are absolutely unlimited.

If you've never experienced operation via OSCAR, you're in for a tremendous experience. The ability to operated *duplex* fashion (simultaneous transmit and receive using separate rigs on separate bands) is truly unique. You hear your own signal exactly as others

Fig. 1-2. These QSL cards represent the outstanding achievement of Worked All States via OSCAR satellite. The proud owner of this WAS certificate #14 is Mike Smithwick, WA6TUF.

hear it. Additionally, several satellite stations can simultaneously operate from one site on a reliable basis. Such capabilities can prove highly beneficial during emergency or contest activities (Fig. 1-2).

Our amateur satellites are quite unique in nature—they do not recognize social, political or economic barriers. Their communication capabilities inspire international friendship, peace and understanding. This aspect allows the spirit of amateur radio to flourish,

Fig. 1-3. WA6TUF as 3A0JB operating portable/mobile near Monaco. 28 different stations were contacted during these activities. Antennas mounted on a tripod followed the satellite's path, which paralleled the mountain in the background.

Fig. 1-4. One of the prime designers and builders of OSCAR 1, Lance Ginner, K6GSJ, holds the spacecraft under his arm before launch day arrives. The successor, OSCAR 2, was identical in appearance to OSCAR 1 (courtesy Project OSCAR).

and sets it apart from all other similar communications media (Fig. 1-3).

HOW IT ALL BEGAN

Amateur radio officially began its sojourn into the satellite program on Dec. 12, 1961, when a simple 10 pound package of electronics was placed in orbit around our planet. The package, known as **OSCAR 1,** transmitted information concerning its internal temperature back to earth in the form of a timed CW "HI" on the 2 meter amateur band. The satellite was home constructed by a small group of amateurs in the California area (Fig. 1-4) and its free ride into space was accomplished by replacing a satellite launch rocket's dead weight ballast with the tiny package (Fig. 1-5). The 100 milliwatt transmitter aboard OSCAR 1 lasted approximately three weeks before its internal batteries discharged, and the satellite fell to earth soon thereafter. Consequent reports of OSCAR 1 reception from over 600 amateurs in 25 countries indicated solid interest in amateur satellites, thus plans for another "bird" were begun.

OSCAR 2 was launched from Vandenburg Air Force Base in California on June 2, 1962. This satellite was identical to OSCAR 1, except its transmitter ran slightly higher output power. A simple spring mechanism separated the satellite from the launch vehicle, and another spring swung its antenna into position. OSCAR 2 lasted approximately 2½ weeks befor its non-rechargeable batteries died.

Fig. 1-5. The satellite that started it all—OSCAR 1. This small unit was designed and built by members of the Project OSCAR group organized at Foothills College in Los Altos, California and Lockeed Amateur Radio Club in Sunnyvale, California. OSCAR 1 was carried aloft by the Discoverer 36 rocket (courtesy Project OSCAR).

Fig. 1-6. OSCAR 3, amateur radio's first full communications satellite. The small group of photocells on each side of this craft were provided as battery backup rather than recharging.

Fig. 1-7. OSCAR 4 undergoing construction during 1964. The following year, OSCAR 5 was launched and a rocket malfunction placed it in a highly elliptical orbit. The first U.S. to U.S.S.R. communications resulted from that mishap.

Over 700 reports of reception again proved support for amateur satellites, and additional plans were instigated.

OSCAR 3 was launched on March 9, 1965, and it made history as amateur radio's first full communications satellite (Fig. 1-6). The one watt, 2 meter linear transponder aboard this craft relayed CW, SSB and FM signals from amateurs around the world. Approximately two weeks after this satellite was placed in orbit, its internal batteries died, and the unit ceased to function. Although a bank of solar cells were incorporated in this satellite, they were strictly for battery back-up rather than for recharging. Over 100 amateurs in 15 countries report communicating via this short-lived satellite.

OSCAR 4 was launched from Cape Kennedy on Dec. 21, 1965. A defect of the launch vehicle caused this satellite to be placed in a highly elliptical orbit, however, the first U.S. to Russia communications was provided by this satellite (Fig. 1-7). The craft employed a 2 meter to 70 cm transponder capable of three watts ouput. A failure of the craft's solar cells caused this satellite to die in March, 1966. A significant lesson in satellite range/altitude was acquired through this bird's misfortune. This information will be used in the Phase III era of long distance satellite communications, which is scheduled to begin in early 1980.

OSCAR 5 was built by students at Melbourne University in Australia, and launched from the U.S. Western Test Range on Jan. 23, 1979 (Fig. 1-8). AMSAT completed extensive modifications, testing, launch arrangements, licensing and data collection for this satellite project. This craft featured two elaborate telemetry systems for sending information concerning its health and well being back to earth. The information was used for planning future OSCAR satellites. One transmitter operated in the 2 meter band, and the other operated on 10 meters. Although this satellite lasted slightly over a month before its batteries discharged, over 200 reports for telemetry reception from amateurs in over 25 countries were received.

Fig. 1-8. A spectacular evening launch carried OSCAR 5 into orbit on Jan. 23, 1970. Performance of this outstanding satellite was something to remember.

Fig. 1-9. OSCAR 6, the classic amateur satellite, mounted on side of Delta Rocket 91 and awaiting launch. The 10 meter antenna was a ribbon-looking folded flexible steel rule which was released when OSCAR was deployed from the rocket.

OSCAR 6, a highly outstanding amateur satellite, was launched from the U.S. Western Test Range on Oct. 15, 1972 (Fig. 1-9). This classic satellite featured a one watt 2 to 10 meter linear transponder which operated beautifully for over four years (Fig. 1-10). Magnetic stabilizing rods were used for accurately controlling satellite roll rate. OSCAR 6 was the first in a series of long-life satellites. Its internal Nicad battery package was continuously recharged by solar cells which covered the craft's outer perimeter. Thousands of reliable QSO's were accomplished by this perfectly operating satellite. Indeed, it seemed that OSCAR 6 could do no wrong. Each time the bird crossed your horizon, the 10 meter satellite band (29,450 to 29,550 kHz) sounded like contest time on

20 meters. The AGC system, which plays an extremely important role in the successful operation of all amateur radio satellites, was perfectly designed and adjusted. Additionally, all satellite users were quite conscientious and cooperative during this era of time. The overall results were that ground stations running low power and simple antenna systems could thoroughly enjoy satellite operations on a daily basis. During 1972, AMSAT received the ARRL Technical Merit Award for their outstanding leadership in the amateur space communications program which resulted in the success of OSCAR 6. Although this satellite is now defunct, its memories will last for many years.

OSCAR 7, the second long-life satellite open to amateurs around the world, was launched on Nov. 15, 1974, from the U.S. Western Test Range (Fig. 1-11). Two rather elaborate transponders and two telemetry/beacon transmitters are included in this spacecraft. The mode "A" transponder is a 2 meter input—10 meter output unit, and the mode "B" transponder is a 70 cm input—2 meter output unit. Both transponders run two watts output. OSCAR 7's electronic subsystems were developed by amateurs in the U.S., Australia, Canada and West Germany. AMSAT provided management and guidance for this project.

This spacecraft's capabilities and sophistication are comparable to the first six OSCAR satellites combined. The array of solar cells covering the satellite's outer perimeters are used for powering its electronics and charging its internal batteries. If the spacecraft is in

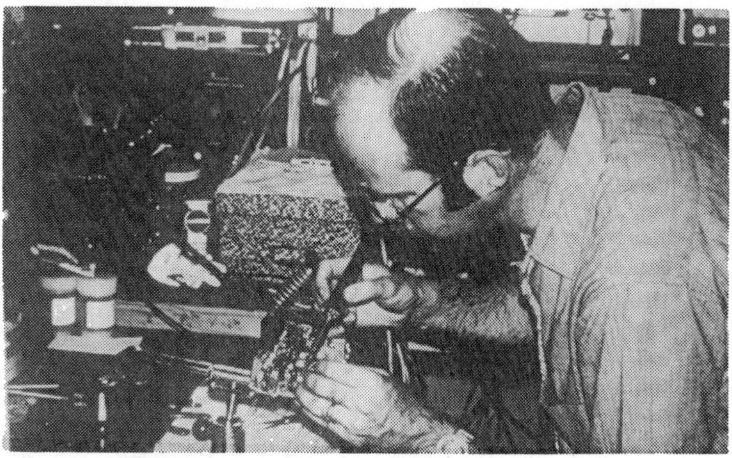

Fig. 1-10. Dick Daniels, WA4DGU, constructing the 2 meter to 10 meter linear transponder used aboard OSCAR 7. Dick built both this repeater and the OSCAR 6 2 meter to 10 meter unit in his basement.

Fig. 1-11. OSCAR 7 lifts off on Delta Rocket 104. This historic event took place on Nov. 15, 1974, and resulted in a perfect orbit for this outstanding amateur satellite.

sunlight, solar cells provide power. If the spacecraft is in darkness, the Nicad batteries provide power. Whenever the battery charge drops 60 per cent below full charge, the satellite switches to recharge mode (mode "C") for approximately 24 hours. At the end of this time, the satellite switches to mode B and continues normal operations. As this information is being written, OSCAR 7 is continuing to perform in an outstanding manner (Fig. 1-12). A slight problem with battery charging and baseplate temperature developed during early 1978, however, a minor change in the satellite's operating schedule appears to have corrected that situation. This problem and its corrective measure does indicate the satellite is showing its first signs of aging. As of mid-1978, OSCAR 7 is operating two days in mode B, and each third day in mode A. A check with recently published tracking data concerning OSCAR 7 will provide additional information on this outstanding amateur satellite.

The AGC system of OSCAR 7 is slightly more active than that of OSCAR 6, which means stations running excessively high power (over 200 watts in the case of OSCAR 7; OSCAR 8's AGC is fully activated at a much lower ERP level) reduce the satellite's sensitivity for other stations. No station should run more than 100 watts ERP. Presently, this situation is also being aggravated by stations

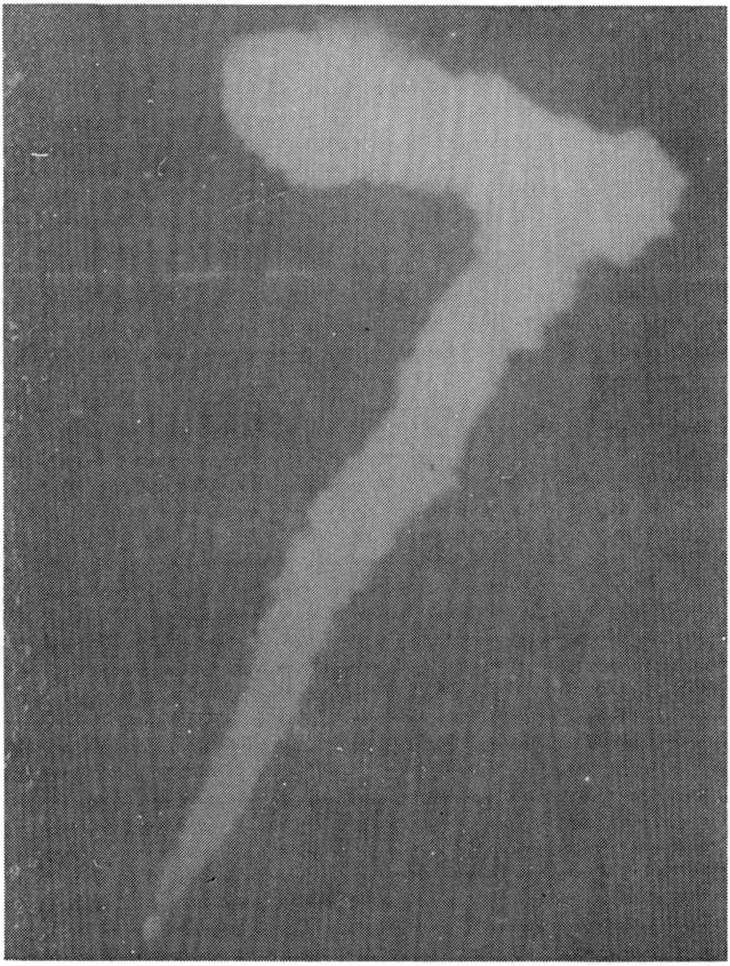

Fig. 1-12. As Delta Rocket 104 rose into the western skies on Nov. 15, 1974, a contrail which developed was caught by a crosswind and formed into a perfect 7. The general reaction was that of good fortune and success for OSCAR 7.

Fig. 1-13. Artist's (WA6TUF) conception of a recently-launched OSCAR 8 in orbit around earth. This craft begain its life in space on March 5, 1978.

running extremely high ERP. Several corrective measures are in the works to eliminate this problem. Additional measures which will be included in Phase III satellites will also discourage high power activities. OSCAR 7 has provided reliable communications for amateurs in many countries, and as of mid-1978 this spacecraft continues doing a magnificent job—long live OSCAR 7.

THE LATEST DEVELOPMENTS

OSCAR 8, our most elaborate satellite to date, was launched from Vandenburg Air Force Base in California on March 5, 1978. This bird was designed and built by amateurs in the U.S., Japan, Germany, and Canada (Fig. 1-13). The flawless launch of this craft produced a perfect orbit (Fig. 1-14), and OSCAR 8 is presently performing beautifully. The slightly lower altitude orbit of OSCAR 8, coupled with its package of new "space ready" electronics, has also resulted in a very sensitive and highly efficient satellite. Two very elaborate transponders and two telemetry/beacon transmitters are in operation aboard OSCAR 8. The mode A transponder is a 2 meter input, 10 meter output unit running approximately two watts. The Japanese-designed mode J transponder is a 2 meter input, 70 cm ouptut unit also running two watts.

Unlike OSCAR 7, the new OSCAR 8 is designed to operate in a positive power format (Fig. 1-15). Basically, this means its outer cover of solar cells and its Nicad battery pack are capable of producing and storing more power than the craft uses. This permits extended periods of operation in the popular mode A. Present calculations indicate that OSCAR 8 should operate for several years, and its orbit can be maintained beyond the year 2050. Barring unforeseen circumstances, this satellite will enjoy a long and prosperous life. OSCAR 8 is presently operating mode A during weekdays, and mode J during weekends. Up-to-the-minute satellite data can be obtained from the monthly amateur radio magazines.

The AGC system of OSCAR 8 is extremely active and highly responsive to all medium strength signals. Likewise, its highly sensitive receiver and low orbit are intended to allow amateurs running low power or simple "bush" communication setups to successfully communicate via OSCAR 8. On the other hand, amateurs running unnecessarily high ERP or a large group of amateurs simultaneously accessing OSCAR 8 will cause its AGC to heavily limit input levels. This results in low power ground stations experiencing difficulty accessing the satellite when power usurpers are present. As many old time satellite users realize, such situations relax after a satellite's newness wears thin. The prime consideration to realize is to stick by your ERP of no more than 100 watts and operate properly. Low power stations (25 to 40 watts ERP) will realize their

Fig. 1-14. Blastoff of OSCAR 8 on the morning of March 5, 1978. The textbook launch was a solid success, and OSCAR 8 acquired perfect orbit (courtesy USAF/NASA).

Fig. 1-15. This gantry photo shows size of OSCAR 8 (right back side of rocket) and Landsat C (top of rocket) satellites. Veteran satellite enthusiast Lance Ginner, K6GSJ, is leaning over the rocket and checking OSCAR's mount (photo by K6PGX).

best results when operating QRP nights. As an example of this, I recently enjoyed five QSO's during a single pass on QRP night while running a mere 12 watts ERP (Fig. 1-16).

AMSAT and its coordinated groups in several countries were responsible for OSCAR 8's construction activities and its placement in orbit. The American Radio Relay League has now assumed the administration and operational responsibilities for OSCAR 8, thus the ARRL is supplying information and technical data to interested parties.

Since OSCAR 8 is presently the star performer in our satellite program, a full description of its operation parameters and capabilities is included in Chapter 3.

PHASE III

AMSAT is presently devoting its full efforts and resources toward completing the first Phase III satellite scheduled for launch in 1980. Considerable progress has been made. This fantastic satellite exceeds all previous OSCARs combined in technical sophistication and operation capability, and it provides amateurs with a truly new communications resource. In order to fully realize the impact of Phase III, a brief overview of the previous amateur satellites is in order. *Phase I* satellites (OSCAR 1 through 5) were short-life, low altitude "get acquainted" units. *Phase II* satellites (OSCAR 6, 7, and 8) are low altitude, long-life, full-communication satellites capable of relaying signals up to 5,000 miles for periods of 20 to 25 minutes at a time.

Phase III satellites, in comparison, are long-life, high-altitude elliptical orbit crafts capable of serving the complete northern hemisphere for 10 to 15 hours each day. In addition to providing efficient use of our VHF amateur bands for long distance communications, these satellites provide the following outstanding abilities:

- The equivalent of a new amateur band with the range capability of 20 meters, the reliability of 2 meters, and the traffic handling ability of 80 meters.
- A reliable means of transmitting specific information throughout entire continents.

Fig. 1-16. Dave Ingram, K4TWJ, with setup used to demonstrate OSCAR communications to news media during 1978. Equipment consists of 432 MHz receiver, 2 meter transceiver with small linear amplifier, 10 meter receiver, and 20 meter transceiver for activity coordination. Television at right was for SSTV experiments via OSCAR 8.

Table 1-1. Preliminary Information on the Russian Satellites.

Period (Time of each orbit)	120.4 minutes
Incrementation (moves West on each orbit)	30.0 degrees
Inclination (angular travel)	82.6 degrees
Beacon Signal	29,400 kHz
Uplink (approximate)	145.900 mHz
Downlink (approximate)	29,380 mHz

- A transcontinental communication medium which may be reliably used for many hours daily by amateurs using relatively simple, low power setups.

The transponder included in the first Phase III satellite is a mode B unit operating 70 cm input and 2 meters output. The transponder runs approximately 50 watts to directional antennas which increase ERP to between 500 and 1,000 watts. A group of three adjacent frequency IF amplifiers, each with its own AGC system, prevent power usurpers or overcrowding conditions from loading the satellite's full output. Additionally, a strong signal seeking *notcher circuit* is incorporated in this arrangement to handle non-complying satellite users. An onboard microcomputer handles housekeeping functions of the craft. Additional information on Phase III is found in Chapter 3.

THE RUSSIAN SATELLITES

The first Russian amateur radio satellites, RS 1 and RS 2, (Radio Sputniks) were launched on Oct. 26, 1978. These satellites performed beautifully (Fig. 1-17). Each satellite features a 2 meter to 10 meter transponder with a beacon signal on 29,400 kHz. The beacons relay telemetry information in the form of four groups of letters and numbers followed by their ID, RS, at 25 words per minute. Both satellites R1 and R2 are in orbits similar to OSCAR 7 and OSCAR 8, except their directions (ascending and descending modes) are reversed. These tiny crafts are approximately 1700 kilometers high. Preliminary reports on RS indicate the satellites have a period of 120.4 minutes; an inclination of 82.6 degrees; and an incrementation (moves west on each orbit) of 30.1 degrees. The beacon signal is 29,400 kHz. The uplink bandpass is approximately

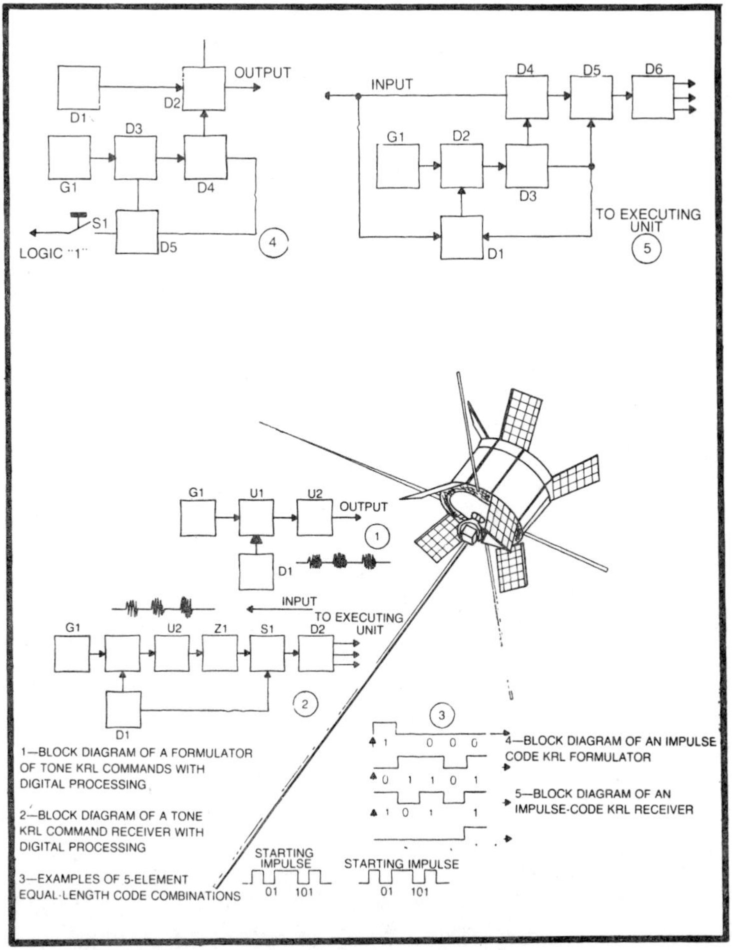

Fig. 1-17. Artist's concept of the Russian satellite in space.

145.880 MHz to 145.920 MHz. The downlink bandpass is approximately 29.360 to 29.400 MHz (Table 1-1).

Satellite users are urged to limit uplink power to 10 watts ERP, otherwise the satellite will overload and shutoff for the remainder of that particular orbit. Additional information on these outstanding satellites may be heard Sundays at 0130 on 7165 kHz during the DX program from Radio Moscow, and via official bulletins on W1AW.

Chapter 2 The Fun of Satellite Operations

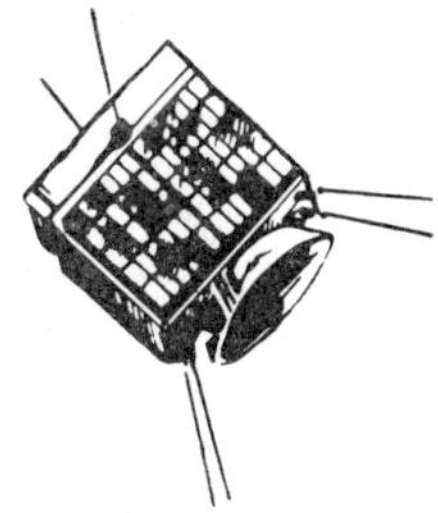

One of the most effective means of describing amateur satellite operations is by relating a typical station's activities during an OSCAR pass. The following armchair visit describes that activity.

Initially, the satellite operator uses information published each month in various amateur radio magazines to determine when specific OSCAR passes are within range of his location. Then, when OSCAR crosses the Equator or approaches his country on such a pass, the satellite gear is switched on and its beacon frequency is monitored. Soon, the beacon's signals are received and activity begins. A frequency relation chart is used to determine approximate in/out frequencies for the particular satellite bands being used. Then, a series of dits or V's are transmitted while the station receiver is tuned to locate these signals as they are returned via the satellite. An antenna direction schedule is often used during satellite passes also, and the operator simultaneously repositions his antennas to correspond with this schedule. Once the satellite operator acquires his return signal, he can start communicating with others through the satellite. Most contacts via OSCAR are relatively short and explicit to permit maximum operating time during a pass. Since all transmitting stations can hear their own signals being returned "duplex style" via the satellite, full break in operation without information repetition is quite feasible.

If his satellite-return signal becomes stronger than other satellite signals, the operator should reduce power or turn his beam off the satellite to avoid overloading the transponder. This courteous maneuver (which may be required as OSCAR passes approximately

30 degrees beyond overhead) assures all satellite users an equal chance to enjoy space communications, while also indicating good operating techniques by the transmitting station.

Then, as the tiny satellite progresses toward the far horizon and signals begin to weaken, the satellite operator brings all his assets into action and searches for DX. Seconds become minutes during these times, since many foreign stations have narrow "windows" of communications with certain areas. The action becomes frantic as the operator calls various stations or changes frequencies to effectively use each precious minute. Frequency drifts caused by doppler shift also give an illusion of actually riding a spacecraft orbiting the earth. Then, just as swiftly as the satellite and its signals appeared, they fade into the noise. The band again becomes "dead" until the next satellite pass begins and the action resumes.

UNLIMITED POSSIBILITIES

The many possibilities associated with satellite operations allow it to perfectly fit the desires of practically every amateur. Let's consider some of these possibilities on an informative basis, and hopefully determine the aspects you find particularly appealing.

Amateurs with limited time available to enjoy our exciting hobby will truly appreciate the precisely scheduled operations of OSCAR satellites. The need to wait for band openings into specific areas is condensed to simple calculations which may be performed hours or days before a specific pass. Likewise, the relatively brief time periods associated with each pass usually can be worked into one's routine when regular low band operations would be impossible. New satellite operators often are pleasantly surprised to find occasional OSCAR passes occuring during lunch times (yes, fixed mobile OSCAR work is quite practical) or within a few minutes after their arrival home for the evening (Fig. 2-1). The pleasure of catching a satellite pass before a busy night's activities get underway has proven very satisfying to many amateurs.

The equipment which may be used for OSCAR communication ranges from inexpensive units to highly sophisticated gear. Thus the satellite enthusiast has a wide variety of options available to match his often varying budget.

CW operation on mode A is the least expensive means of getting involved with OSCAR activities. A 145 MHz transmitter is the only additional gear one must add to his high frequency setup for this operation. Two meter FM rigs or surplus high band government radios also may be temporarily pressed into OSCAR use by properly keying their final amplifier stage and disconnecting the microphone. VFO problems with many high frequency transmitters are often "side-stepped" by the use of variable slug tuned coils or trimmer capacitors placed in series or parallel with the unit's crystals.

While exclusive CW operation via OSCAR is quite rewarding and highly successful, most satellite enthusiasts also desire SSB capabilities for use when strong signal conditions prevail. There are several multimode UHF transceivers presently available which

Fig. 2-1. OSCAR moble/portable setup of WA6TUF as used in Europe during 1977. Equipment includes ICOM IC-211 for mode B reception, KLM Echo 70 and 30-watt amplifier for 432 MHz uplink, 4-element beam antenna for 2 meters and 10-element beam antenna for 70 cm. The antennas are placed on a camera tripod placed beside the car during satellite passes. Manual satellite tracking techniques are used. Note accessibility of mike and key for instant mode change.

serve this purpose beautifully. Such rigs are manufactured by companies like ICOM, KLM, Yaesu, and Kenwood. Many of these rigs feature digital readout dials, semi-break in CW options, auxiliary VFO's, automatic noise limiters, etc. The cost of these units range from $800 to $1200. Two less expensive transceivers which are particularly appealing for OSCAR use are the ICOM 202S, 2 meter unit and ICOM 402 70 cm unit. These small transceivers may be operated fixed, portable, or mobile as the occasion requires. Their receiver sensitivity is very good, and transmitter output power is approximately three watts. If a small linear amplifier is included with these units, a compact and efficient OSCAR station is easily produced.

Amateur satellite setups planned around small transceivers like the ICOM units are particularly appealing for portable and mobile operations (Fig. 2-2). The OSCAR users desiring to enjoy brief activities during weekend jaunts, for example, merely packs his condensed OSCAR package along with the usual camping or fishing gear, and he's set. If a small metal or wooden framework is used so it may be moved from room to room in the house, used mobile on the car seat, or operated portable without the usual entanglements of cabling and disconnecting specific units for use.

Another unique feature of satellite activity is the capability of full duplex operations. This ability to hear your own signals being returned by OSCAR as others hear them is quite fascinating and useful. The time lag, coupled with satellite doppler shift is unlike anything the amateur may experience elsewhere. The "self reception" aspect also is highly useful, for equipment or antenna experimentation and evaluation. Quite simply, when your hear your signals being relayed by OSCAR, you know exactly when you can contact the other station—or why your calls fail.

The reduced size of antennas used for satellite work are particularly appealing to many amateurs. Such antennas usually minimize neighborhood TVI complaints, and are often the ideal solution for cliff dwellers. Likewise, small experimental antennas are easily constructed and tested via OSCAR.

FOLLOWING A BIRD

One of the more interesting and effective ways of being involved with satellite communications is by following a bird from its initial development, design and blast-off phases, into its placement in orbit, and then sharing in its use throughout its lifetime. The most recent example of this situation is our latest satellite, OSCAR 8. AOD, as it was termed before launch, was constructed primarily by

Fig. 2-2. WA6TUF/4U1 operating portable/mobile via OSCAR from the International Telecommunications Union parking lot in Geneva, Switzerland. Equipment was simply placed on the car seat for duration of a pass. Combination of 70 cm and 2 meter twist antennas were mounted on camera-type tripod and hand rotated during a pass.

amateur satellite enthusiasts in Japan and America. When construction of this small unit was completed and a "piggyback" ride on Delta Rocket 139 was secured, the satellite was strapped into a passenger seat on a commercial airlines and flown to California for launch from our landsat C site at Vandenburg Air Force Base (Fig. 2-3).

The textbook-type launch on March 5, 1978 went perfectly, with AMSAT Network stations operating on 14,290 kHz providing a second by second account of each orbital correction burn or maneuver. Approximately one hour and a half after OSCAR 8 was lifted from its launch pad in California, the announcement rang around the world via 20 meters, "Gentlemen we have a new green bird in perfect orbit." Thus, our package of amateur-built electronics, approximately the size of the Drake L4B linear, became the latest satellite in orbit around earth (Fig. 2-4). That same evening, several amateurs and myself acquired visual contact of the satellite on the eastern horizon. During the next pass of OSCAR 8 it was determined that satellite roll rate was extremely low and the bird's 10 meter antenna was deployed. During the next pass, the satellite's 2 to 10 meter transponder was activated and successful operation of the complete satellite was confirmed.

On the air enthusiasm ran high during times as control stations described their operations via 20 and 80 meters while the satellite responded accordingly on 10 meters. Amateurs were thus afforded a ring side seat as the antenna's drive motors whirred into action and the transponders beacon was switched into operation. One could actually feel the tension and sighs of relief as each "no-backup" step was executed. Now, OSCAR 8 has been in orbit for many months and it continues to perform beautifully, giving amateurs around the world their own communications satellite which can be enjoyed on a daily basis.

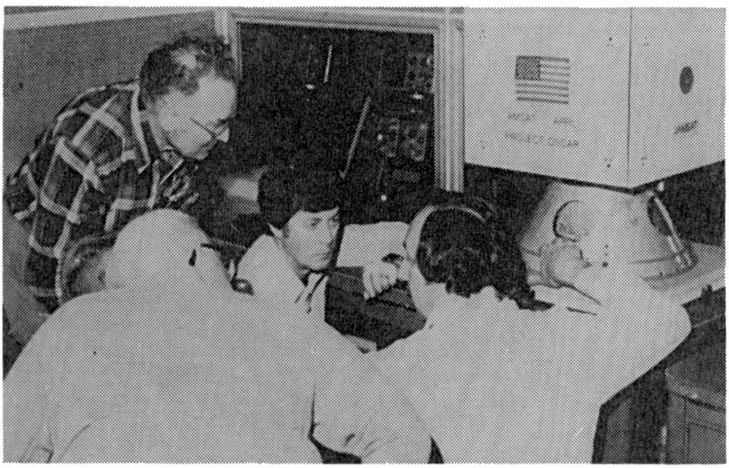

Fig. 2-3. Following a thorough checkout of Vandenburg Air Force Base, satellite enthusiasts K6PGX, W9KDR, W6PAJ, W6CG and W6ELT (left to right) gather around OSCAR 8 and discuss where the conical mount will be removed and replaced with the launch rocket mount. This photo was taken within a few hours of the satellite's journey into space.

Fig. 2-4. Jan King, W3GEY, checks out the marmon clamp on OSCAR 8 to assure its accurate operation during ejection from the launch vehicle.

The world of satellite communciation does by no means dwindle at this time. Indeed, the best is yet to come! Phase III, our next group of amateur satellites, opens a complete new world of space communications for all amateurs. These satellites are high altitude spacecrafts, providing almost worldwide communications for periods of several hours each day. One of the unique capabilities of such a "high flying bird" is simultaneous direct communications in all areas of the northern hemisphere.

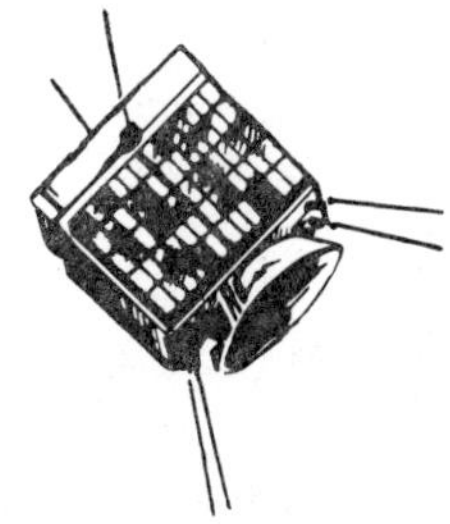

Chapter 3
The Satellites

The activities associated with designing, constructing and placing a communications satellite in orbit around our world are far more complex than most amateurs realize. Indeed, a concise discussion of these activities is far beyond the scope of this book. We will, however, look at a brief synopsis of some basic considerations.

INSIDE THE SATELLITES

In order to assure reliable operation of any space satellite, each component must be highly reliable and "space ready" quality (Fig. 3-1). Simple resistors, capacitors, etc. meeting these high standards usually cost 10 to 15 times more than conventional components. Likewise, "space ready" transistors and IC's which are produced for extreme use and long lifetimes are quite costly but necessary for our satellites (Fig. 3-2).

Special Nicad batteries, which must pass stringent tests under extremes of environmental conditions are hand picked for use in each satellite. Their wiring configurations are likewise vitally important to sustain extended use without malfunctions (Fig. 3-3).

Solar cells which cover a satellite's outer perimeters (Fig. 3-4) also must be space ready quality, and each one must be protected from the sun's radiation by a special lens. Such solar cells, screened to NASA standards are quite expensive.

Many services, like special machine work and space ready wiring techniques, must be acquired for each amateur satellite launched.

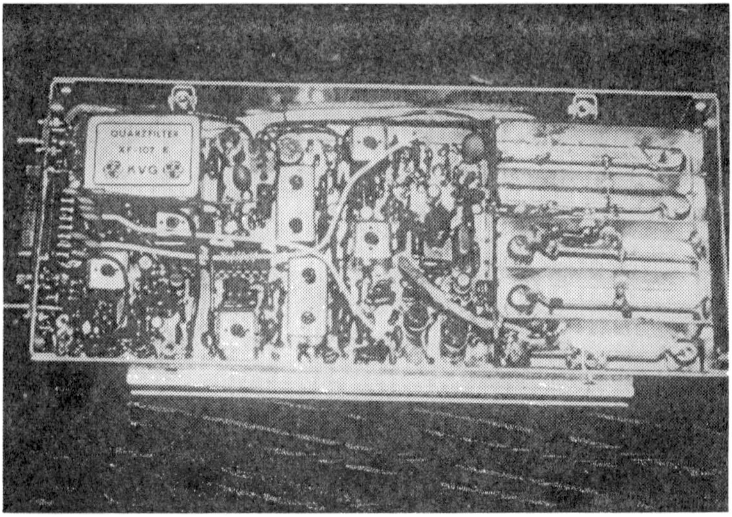

Fig. 3-1. 70 cm to 2 meter transponder receiver used aboard OSCAR 7. This unit was built by DJ4ZC, Karl Meinzer, and DJ5KQ, Werner Haas.

Accommodations for each satellite's ride into space must be secured, plus administration and management responsibilities must be assumed.

Think of these parameters the next time you hear an OSCAR satellite and you, too, will agree these crafts are not simply "adult toys."

Fig. 3-2. Matching 70 cm to 2 meter transponder transmitter used with OSCAR 7. Note space ready quality and components.

Fig. 3-3. Interior view of OSCAR 7, before solar cell panels were installed. The two transponders, 435 and 2304 beacons comprise the top row of modules. The second row contains the teletype telemetry encoder, command decoders, a 28 volt regulator, and the battery charge regulators.

FUNDING THE SATELLITES

In the same manner that amateur FM repeaters are supported by funds and donations from their users, amateurs satellites are funded by contributions from our amateur fraternity. While such donations are not mandatory for satellite use, they should be considered the gentlemanly procedure for each satellite enthusiast. A donation of $10 each year pays for one solar cell on a satellite. This single cell doesn't come close to approaching the yearly energy level used by one individual operator, however, the technique somehow continues to support our amateur satellite program. Amazing!

AMSAT, the Radio Amateur Satellite Coporation, at P.O. Box 27, Washington, DC, 20044, collects OSCAR funds and coordinates the activities. In the same way the ARRL has been established as a headquarters for amateur activity in the U.S., AMSAT is established as headquarters for amateur OSCAR satellites. An AMSAT QSL bureau is also provided for satellite users, and life membership is available for all satellite enthusiasts (Fig. 3-5).

QSL BUREAU

The AMSAT QSL bureau was established as a service to help distribute QSL cards for contacts via OSCAR satellites (Figs. 3-6 and 3-7). WB2LCC is presently handling this operation. The bureau's address is good at his callbook address or via AMSAT. In order to use the OSCAR QSL bureau, one merely sends his cards for distribution, and sends some self-addressed stamped envelopes (SASE's) of his own for the WB2LCC file. Non U.S. stations should send SAE's and IRC's for the file. Cards are exchanged between North American stations at no charge (provided, of course, each has

Fig. 3-4. OSCAR 7 with solar panels in place. The panels are made of 1 by 2-cm N-on-P solar cells. The eight-sided structure is 90 per cent covered with solar cells, sufficient to provide in excess of 15 watts when illuminated by the sun. The spacecraft is mounted on the Delta 104 Rocket under the NOAA 4 meteorological satellite. An additional "passenger" not visible in this picture is the Spanish-built Intasat craft.

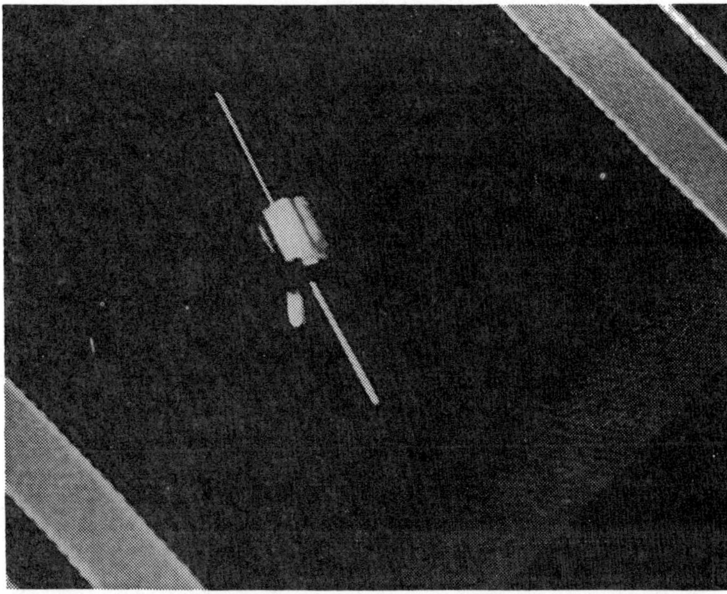

Fig. 3-5. AMSAT OSCAR lapel/tie pin is a gold mini-replica of OSCAR 7 awarded to all life members of AMSAT.

an SASE on file!) from the bureau. DX cards are forwarded at the rate of 6 cents per card. DX stations may send QSLs to bureau users at no charge. The OSCAR QSL bureau is a non profit plan which can be used by all satellite enthusiasts, provided they have SASE's on

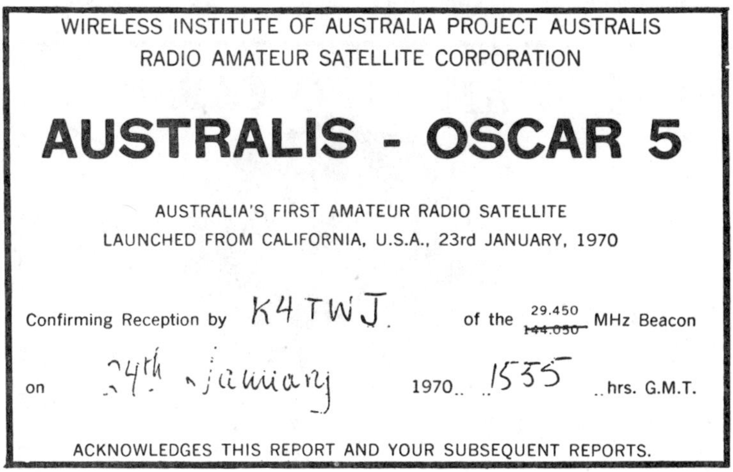

WIRELESS INSTITUTE OF AUSTRALIA PROJECT AUSTRALIS
RADIO AMATEUR SATELLITE CORPORATION

AUSTRALIS - OSCAR 5

AUSTRALIA'S FIRST AMATEUR RADIO SATELLITE
LAUNCHED FROM CALIFORNIA, U.S.A., 23rd JANUARY, 1970

Confirming Reception by K4TWJ of the 29.450 ~~144.050~~ MHz Beacon

on 24th January 1970 1555 hrs. G.M.T.

ACKNOWLEDGES THIS REPORT AND YOUR SUBSEQUENT REPORTS.

Fig. 3-6. QSL card from the now-famed OSCAR 5 satellite. AMSAT distributed these QSLs to amateurs reporting accurate reception of the satellite's telemetry information.

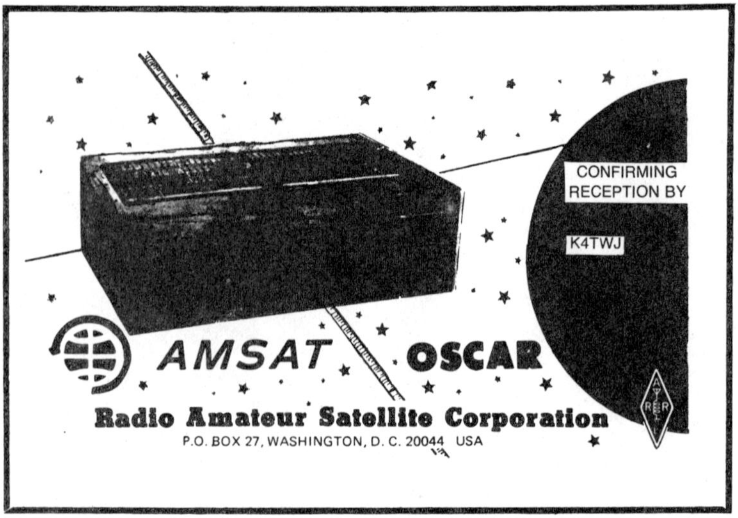

Fig. 3-7. QSL card from the classis OSCAR 6 satellite. This craft featured a 2 to 10 meter transponder which created history during its long lifetime.

file with WB2LCC to receive their QSL's. Additional information is available directly from the bureau. SASE, please (Figs. 3-8 and 3-9). AMSAT headquarters address is P.O. Box 27, Washington, D.C. 20044.

OSCAR 6

Although now retired, this satellite operated in a useful and professional manner throughout its lifetime. OSCAR 6 carried a 2 to

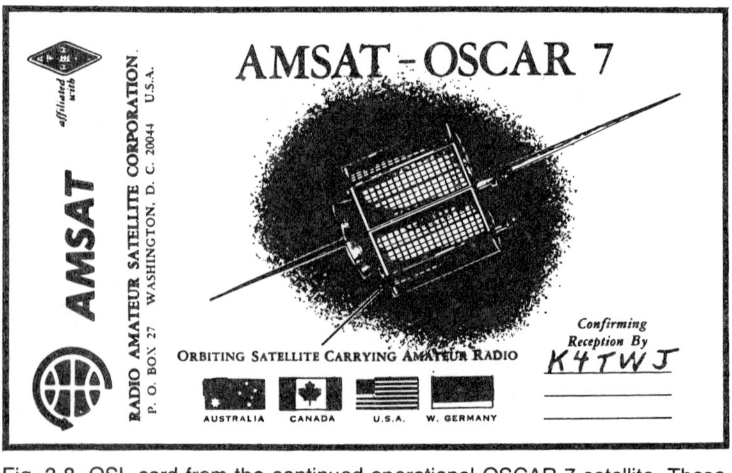

Fig. 3-8. QSL card from the continued-operational OSCAR 7 satellite. These QSLs are distributed by AMSAT.

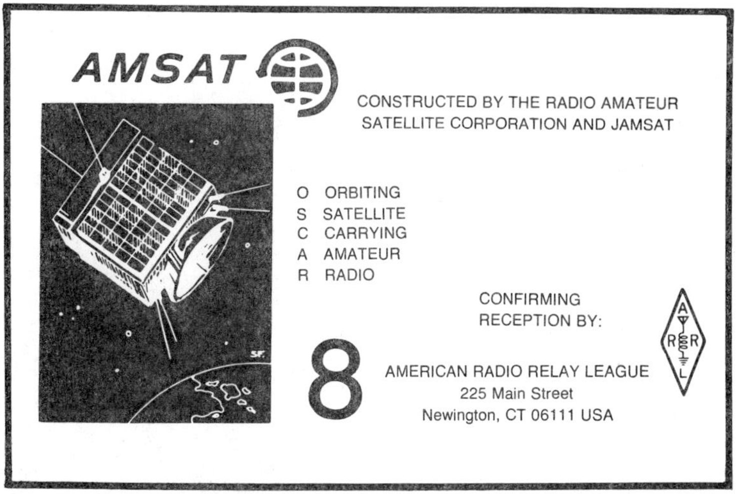

Fig. 3-9. QSL from OSCAR 8, the newest and most sophisticated amateur satellite presently orbiting the earth.

10 meter transponder which had a very sensitive receiver, solid output transmitter and perfectly adjusted AGC. The craft's reliability was beyond reproach, and it truly set the pace for Phase II of the OSCAR satellite program.

OSCAR 7

This satellite is the second craft in the Phase II series of long life satellites. As of early 1980, OSCAR 7 continues operating (Table 3-1). Two transponders are incorporated in OSCAR 7: a 70 cm to 2 meter mode B unit, and a 2 meter to 10 meter mode A unit. Since the AGC system of OSCAR 7's mode A is slightly more active than that of OSCAR 6 and because the craft's batteries recently have shown their first sign of aging, OSCAR 7 presently is spending the majority of its time in mode B (see Table 3-2). DX capabilities of this mode are very good, and downlink signals are relatively strong. One can expect good results via mode B when using approximately 100 watts ERP on the 432 MHz uplink, and an all mode 2 meter transceiver for the downlink.

If you use one of the popular "twist" antennas for uplink satellite communication and you live in the northern hemisphere, it should be phased for left hand circular polarization for mode A and right hand circular polarization for mode B. Such phasing instructions and suggestions for in-shack phase switching arrangements are included with commercial arrays and in many antenna handbooks. Downlink

Table 3-1. Telemetry Information for OSCAR 7.

CHANNEL	MEASURED PARAMETER	MEASUREMENT RANGE	CALIBRATION EQUATION (Prelim.)
1A	Total Solar Array Cur.	0 to 3000 ma	$I_T = 29.5\ N$ (ma)
1B	+X Solar Panel Cur.	0 to 2000 ma	$I_{+x} = 1970 - 20\ N$ (ma)
1C	−X Solar Panel Cur.	0 to 2000 ma	$I_{-x} = 1970 - 20\ N$ (ma)
1D	+Y Solar Panel Cur.	0 to 2000 ma	$I_{+y} = 1970 - 20\ N$ (ma)
2A	−Y Solar Panel Cur.	0 to 2000 ma	$I_{-y} = 1970 - 20\ N$ (ma)
2B	RF Pwr. Out 70/2	0 to 8 watts	$P_{70/20} = 8(1 - 0.01\ N)^2$ (watts)
2C	24-Hr. Clock Time	0 to 1440 min.	$t = 15.16\ N$ (min) or $0.253\ N$ (hrs)
2D	Batt Chg/Dischg Cur.	−2000 to +2000 ma	$I_B = 40\ (N - 50)$ (ma)
3A	Battery Voltage	6.5 to 16.4 V.	$V_B = 0.1\ N + 6.4$ (volts)
3B	Half-Batt Voltage	0 to 10 V.	$V_{1/2B} = 0.10\ N$ (volts)
3C	Bat. Chg. Reg. #1	0 to 15 V.	$V_{crl} = 0.15\ N$ (volts)
3D	Battery Temperature	−30 to +50°C	$T_{Bat} = 95.8 - 1.48\ N$ (°C)
4A	Baseplate Temp.	−30 to +50°C	$T_{bp} = 95.8 - 1.48\ N$ (°C)
4B	PA Temp.-2/10 Rptr	−30 to +50°C	$T_{10} = 95.8 - 1.48\ N$ (°C)
4C	+X Facet Temp.	−30 to +50°C	$T_{+x} = 95.8 - 1.48\ N$ (°C)
4D	+Z Facet Temp.	−30 to +50°C	$T_{+z} = 95.8 - 1.48\ N$ (°C)
5A	PA Temp.-70/2 Rptr	−30 to +50°C	$T_2 = 95.8 - 1.48\ N$ (²C)
5B	PA Emit. Cur. 2/10	0 to 1167 ma	$I_{10} = 11.67\ N$ (ma)
5C	Modul. Temp. 70/2	−30 to +50°C	$T_m = 95.8 - 1.48\ N$ (²C)
5D	Instr. Sw. Reg. Input Cur.	0 to 93 ma	$I_{isr} = 11 + 0.82\ N$ (ma)
6A	RF Pwr Out-2/10	0 to 10,000 mw	$P_{2/10} = \frac{N^2}{1.56}$ (mw)
6B	RF Pwr Out-435	0 to 1,000 mw	$P_{435} = 0.1(N^2) + 35$ (mw)
6C	RF Pwr Out-2304	0 to 100 mw	$P_{2304} = 0.041\ N^2$ (mw)
6D	Midrange Telemetry Calibration	0.500 V.	$V_{cal} = 0.01\ N\ (0.50 \pm 0.01)$

Similar informaiton for OSCAR 8 is available from The American Radio Realy League

signals from OSCAR 7 are linearly polarized, making their reception easy (see Table 3-3).

OSCAR 8

This newest amateur satellite is the third Phase II spacecraft to be placed in orbit around our earth. The craft presently is performing perfectly, and all indications point toward a long and healthy life for OSCAR 8 (Fig. 3-10).

There are two active transponders aboard OSCAR 8: A 2 to 10 meter mode A unit,and a Japanese developed (JAMSAT) 2 meter to 70 cm mode J unit (Table 3-4). Power output of either transponder is approximately 2 watts. Amateurs in the northern hemisphere using "twist" antennas for uplink signals should use left hand circular polarization for mode A and right hand circular polarization for mode J. The satellite's downlink signals are linearly polarized, making their reception with various antennas a simple task. There is a beacon for each transponder aboard OSCAR 8, and each sends a variety of operational parameters to ground-based amateurs (Table 3-5). Details on telemetry decoding information is available from the American Radio Relay League, 225 Main St., Newington, Connecticut

Table 3-2. OSCAR 7 Statistics.

MODE A	145.850 MHz to 146.950 MHz uplink, 29.400 to 29.500 MHz downlink. Output is 1-2 watts. Telemetry beacon on 29.502 MHz and 435.103 MHz.
MODE B	432.125 MHz to 432.175 MHz uplink, 145.925 MHz to 145.975 MHz downlink. Output is approximately 2 watts low, 10 watts high. This transponder inverts frequencies: uplink signals on USB become downlink signals on LSB, etc. Telemetry beacon on 145.972 MHz.
MODE C	Low power equivalant to Mode B. Same 70 cm and 2 meter frequencies utilized, but transponder power is 1 to 2 watts.
MODE D	Recharge mode. Both transponders off.
ADDITIONAL	A 2304 MHz beacon transmitter is also included in OSCAR 7, however the FCC has not authorized its use at the present time.
PERIOD	OSCAR 7 requires 114.94 minutes for one complete revolution of the Earth.
INCREMENT	OSCAR 7 progresses 28.736 degrees west for each orbit.
ALTITUDE	910 miles

Table 3-3. OSCAR 7 Operating Frequency Relations.

MODE A

Downlink frequency = uplink frequency (MHz) – 116.450 (± Doppler shift)

Uplink (MHz)		**Downlink** (MHz)
145.855	=	29.405
145.865	=	29.415
145.875	=	29.425
145.885	=	29.435
145.895	=	29.445
145.905	=	29.455
145.915	=	29.465
145.945	=	29.495
145.925	=	29.475
145.935	=	29.485
145.945	=	29.495

MODE B

Downlink Frequency = note frequency inversion

Uplink (MHz)		**Downlink** (MHz)
432.125	=	145.975
432.135	=	145.965
432.145	=	145.955
432.155	=	145.945
432.165	=	145.935
432.175	=	145.925

06111. OSCAR 8 is presently operating mode A on weekdays (Wednesdays, UTC, are reserved for experimental use only), and mode J on weekends. OSCAR bulletins and operational schedules, which are published in the monthly amateur magazines, should be consulted for additional late-breaking information.

OSCAR 8 is highly sensitive to signals from ground-based amateurs (Fig. 3-11). A relatively low power signal will fully access the satellite, and additional uplink power causes the AGC to react and limit other low power amateurs trying to access the transponder. The results as of late 1978 are less than optimum, as a group of amateurs running high power (400 to 1000 watts ERP) can effectively *desense* the satellite. Since we can't physically reach out and adjust the satellite, this correction must be handled from a management standpoint (Fig. 3-12). The ARRL and AMSAT are thus insisting all OSCAR 8 users employ low power, and conscientiously monitor their downlink signal to assure it is weaker than the beacon signal. Consideration for others and polite manners are absolute necessities for long term success of OSCAR 8. If this policy is followed, OSCAR 8 could rival the classic satellite, OSCAR 6, in success.

Operating OSCAR 8 is easy and enjoyable. The quickest and least expensive way of getting on OSCAR is to use a surplus 2 meter

Fig. 3-10. OSCAR 8, sporting its protective panels and awaiting final checkout at Vandenburg's laboratories.

Table 3-4. OSCAR 8 Statistics.

MODE A	145.850 MHz to 145.950 MHz uplink, 29.400 MHz to 29.500 MHz downlink. Output is 1-2 watts. Telemetry beacon on 29.402 MHz.
MODE J	145.900 MHz to 146.000 MHz uplink, 435.100 MHz through 435.200 MHz downlink. Output is 2 watts. Telemetry beacon on 435.095 MHz. Uplink signals are inverted.
MODE D	Recharge mode. Both transponders off.
PERIOD	OSCAR 8 requires 103.228 minutes for one complete revolution of the Earth.
INCREMENT	OSCAR 8 progresses 25.808 degrees west for each orbit.
ALTITUDE	560 miles

rig for uplink and your existing station receiver for the 10 meter downlink. Suitable 2 meter rigs range from discarded taxicab units to CW-modified FM rigs or 2 meter multimode transceivers. Portable 2 meter sideband rings are also perfect for satellite use. Operation of OSCAR 8's mode J will require the additional capability of receiving SSB/CW signals on the amateur 70 cm band. Several manufacturers distribute suitable 70 cm to 10 meter converters,and at least two manufacturers distribute 70 cm transceivers (Table 3-6).

PHASE III

Our Phase III series of amateur satellites promises new communications horizons for all radio amateurs.

Basically, the concept of Phase III involves the use of high altitude, elliptical orbits and highly sophisticated satellites of the *active* variety. The term active means the satellites actually contain their own thrust motor which can be used to alter or vary paths after initially being placed in orbit (Fig. 3-13). High altitude, elliptical orbits cause a Phase III satellite to briefly circle the earth's southern hemisphere, then spend ten to fifteen hours traveling 10 to 20 thousand miles into space above the earth's northern hemisphere. During this "high altitude time," all amateurs in the northern hemisphere are able to simultaneously communicate via the satellite. Such communications capabilities have never been possible in amateur radio! As an example of Phase III communications, a round table

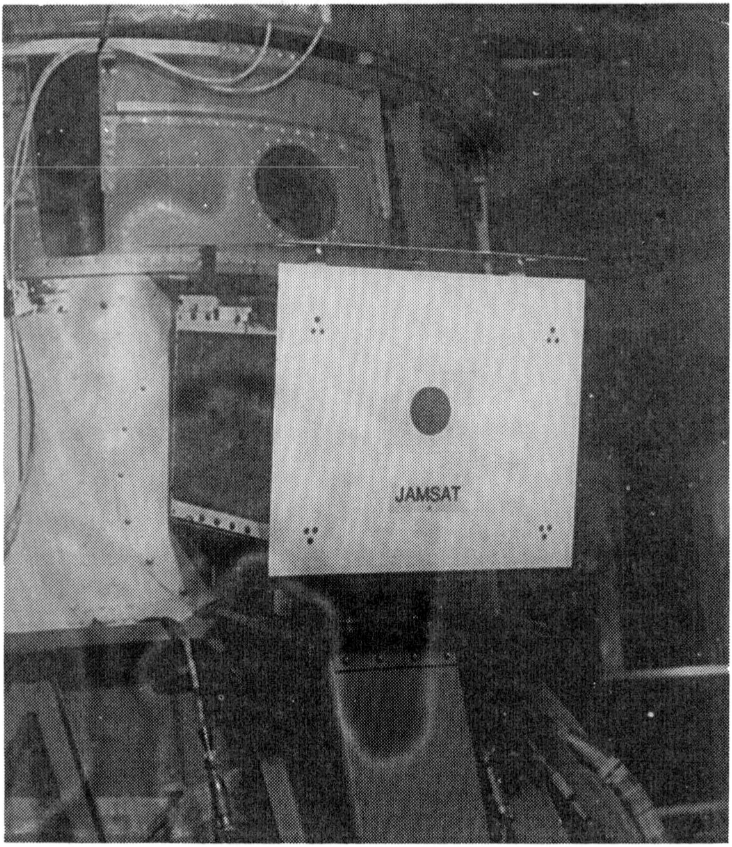

Fig. 3-11. OSCAR 8, being readied for launch is shown attached to its Delta Rocket (photo by K6PGX).

discussion may include two U.S. amateurs a few blocks apart in California, a New England based amateur, some Japanese stations and one or two European stations. Band conditions will not be adversely affected by weather conditions, and full duplex operations will be commonplace.

Since 10 meter signals will not penetrate the upper levels of our earth's atmosphere without severe attentuation, this band is deleted from Phase III use. Alternately, the 2 meter and 70 cm amateur bands are employed because of their atmosphere-penetrating effect (Table 3-7).

Phase III A is progressing very well and plans for a triple-split band system, each with its own AGC similar to that used on OSCAR 6, and an excessive-power uplink filter may be employed. Mode B operation is used, and tentative frequency ranges are as follows:

- **Uplink:** 435.150 MHz to 435.290 MHz.
- **Downlink:** 145.850 to 145.990 MHz.

The beacon frequency is 145.995 MHz. High gain antennas are used to produce an effective radiated power of 500 to 1000 watts from the satellite. Another satellite (Phase III B) is for use during the mid 1980's.

Phase III A rides into space aboard an ARIANE launch vehicle which is a producer of the European Space Agency. On March 22, 1978 the European Agency (ESA) conducted a payload interface meeting which AMSAT attended. The meeting was held in Toulouse, France and Jan King, W3GEY and Karl Meinzer, DJ4ZC attended. Many of the detailed interfaces of the ARIANE vehicle were discussed and many safety items relating to kick motor were reviewed in detail. AMSAT's scheme for firing the motor and for keeping it safe while on the launch vehicle were accepted by ESA.

Table 3-[illegible] OSCAR 8 Operating Frequency Relations.

MODE A

Downlink Frequency = Uplink Frequency (MHz) − 116.458 (± Doppler shift)

Uplink (MHz)		**Downlink** (MHz)
145.860	=	29.402
145.870	=	29.412
145.880	=	29.422
145.890	=	29.432
145.900	=	29.442
145.910	=	29.452
145.920	=	29.462
145.930	=	29.472
145.940	=	29.482

MODE J

Downlink Frequency = 581.1 − Uplink Frequency (± Doppler shift)
Note frequency inversion

Uplink (MHz)		**Downlink** (MHz)
145.910	=	435.190
145.920	=	435.180
145.930	=	435.170
145.940	=	435.160
145.950	=	435.150
145.960	=	435.140
145.970	=	435.130
145.980	=	435.120
145.990	=	435.110

Table 3-6. OSCAR Quick Reference Chart.

SATELLITE	UPLINK FQ	DOWNLINK FQ	BEACON FQ
OSCAR 7 Mode A	145.850 to 145.950	29.400 to 29.500	29.502
OSCAR 7 Mode B	432.125 to 432.175	145.975 to 145.925	145.972
OSCAR 8 Mode A	145.850 to 145.950	29.400 to 29.500	29.402
OSCAR 8 Mode J	145.900 to 146.000	435.100 to 435.200	435.095
PHASE III Mode B (Two transponders used)	435. MHz 100 to 200 kHz bandwidth. Exact frequencies not established at time of this publication.	145. MHz 100 to 200 kHz bandwidth. Exact frequencies not established at time of this publication.	145. MHz band

Structure

More than any other part of the spacecraft, the structure has undergone an evolutionary process. After a number of preliminary concepts were discarded and two different models were built, a structure knows as the ETU (Engineering Test Unit) was built in West Germany. The structure was then sent to the U.S. where a dummy kick motor and wooden modules were installed to simulate the flight spacecraft. In December 1977 the ETU was sent to the Cal. Tech. Jet Propulsion Lab in California were it was subjected to vibration qualification tests at levels specified for the ARIANE vehicle. While the structure passed the tests successfully, a number of changes were suggested by the test results. These changes were incorporated into the final design. In the meantime, the ETU was returned to AMSAT-DL for inclusion of engineering model electronic modules. The completed ETU was then to be sent to SAE for further testing.

Table 3-7. Universal Satellite Bandplan for Amateur Operations.

5%	30%		30%		30%	5%
GUARD	CW	RTTY		SSTV	SSB	GUARD

Mode A	29.400	29.433	29.467	29.550
Mode B	145.925	145.940	145.955	145.975
Mode J	435.100	435.133	435.167	435.200

IHU

The IHU which contains the COSMAC microprocessor was tested at the prototype level for many thousands of hours. The command detector and telemetry encoder schemes were tested in prototypes with CDP-1801 COSMAC while located at a remote repeater site (60 miles from the control site). The results (in terms of bit error rate) were extremely close to the theoretically predicted performance. Features were added to the IHU to allow range measurements to be made via the command and telemetry links, and to allow retransmission of the results of range measurements from one command station to another, again via the IHU.

Transponder

As this chapter is being written, the 50-watt version of the 70 cm to 2 meter transponder is at AMSAT-DL. Werner Haas, DJ5KQ, completed the design of the front end and IF stages and the power amplifier and modulator stages. One flight-quality transponder was completed by mid 1978. The first Phase III spacecraft (Phase III-A) will not carry both frequency combinations as had originally been hoped for. The 70 cm-to-2 meter transponder was chosen for development first because it provides better link performance. The 2304 MHz S-Band beacon was dropped for the same reason and because no allocation can be assured until the 1979 World Administrative Radio Conference.

One particular problem associated with the transponder has been solved. Until recently it was not possible to find a good quality crystal filter with a bandwidth of 150 kHz. JAMSAT members approached the Japanese firm which provided the filter for the A-O-8 mode J transponder regarding this problem and they provided a 150-kHz bandwidth filter to AMSAT specifications.

Antenna Subsystem

The antenna system is also very critical to proper Phase III performance. With the model designed by Tom Clark, W3IWL, it was possible to determine that one antenna system at the end of the

Fig. 3-12. High above ground in the Delta launch gantry, a final checkout is made to assure everything is ready for the trip into space. Jan King, W3GEY, stands beside OSCAR 8 on the rocket's right side as Lance Ginner, K6GSJ, (kneeling) checks support wiring.

Table 3-8. Cost Breakdown of Phase III Satellites.

2400 solar cells	$10 each	$24,000
12 battery cells	$200 each	2,400
Kick motor		10,000
2 transponders	$5,000 each	10,000
2 Cosmac computers	$8,000 each	16,000
	Total	$62,400

arms could not be used on 2 meters as well as 70 cm. Thus a separate 70 cm antenna was placed along the spin axis of the spacecraft on the end opposite the motor. A one-third scale model of Phase III was recently constructed by Bill Hodzik, WA2UDT, so that detailed antenna pattern measurements could be made. These measurements took place at NASA so that flight antennas could be built from this data.

Power Subsystem

AMSAT worked with NASA and NOAA to obtain 48 battery cells remaining from the ITOS satellite program. In addition, some 10,000 solar cells may also be obtained from the same source. While the cost of the cells is itself high, even if the cells are donated, the cost of the cover slips and fabrication still amount to approximately $10 per cell (Table 3-8).

The battery charge regulator (BCR) for Phase III is very similar to that flying in AMSAT OSCAR 8. Although A-O-8's BCR is less sophisticated than that for Phase III, it verifies the concept of converting power from 28 volt solar arrays to a 12-14 volt battery system which is the same for both satellites. In Phase III, the BCR and all other DC-to-DC converters are contained in a single power module.

Attitude Control Subsystem

The concept that makes Phase III so different from previous satellites is that it has an active attitude control system. This difference is quite profound. While former satellites were mere boxes with electronics inside, this spacecraft is a robot which can be instructed to assume any attitude we choose in space under remote control and it can tell us when it gets there. We have one constraint with our robot, though—it wants to spin about one axis. In fact, with the IHU, we may store nearly any number of motions of the spin axis for execution at a later time or under a certain set of conditions.

The components for this system were selected as follows: One "eye" of the robot is similar to a sun sensor used previously on another NASA mission. This sensor tells the spacecraft where the sun is in relation to the spin axis and when the sun passes by one of the three arms. The second "eye" is an earth sensor which gives the relationship of the earth to the satellite at various points in the orbit. The earth sensor was developed by DJ4ZC with components provided by Leitz, a West German optics manufacturer. The force applied by the satellite to cause it to attain the proper attitude is via interaction of the earth's magnetic field with a larger magnetic torquing coil network aboard the spacecraft. Most of the electronics required in this subsystem is to interface these components with the IHU. Many of the logic operations which were once handled by discrete logic controllers can now be performed by the software. A

Fig. 3-13. Structural model of AMSAT Phase III satellite scheduled for launch during early 1980. This model has undergone a number of simulated flight tests and refinements and is now qualified to mate with its European Space Agency ARIANE launch vehicle.

prototype of the torquer coil assembly and the earth sensor w
assembled in West Germany.

The remaining component in the attitude control system is a fluid damping system. It consists of small tubes partially filled with a viscous fluid, and will stop any nutation or wobble of the spin axis in a matter of seconds. Prototypes of these tubes have been built and tested and are being incorporated into the ETU.

Ground Stations

Ground operations for the Phase III launch are far more demanding than those needed for Phase II satellites. For this reason, ground equipment needs to be finished well ahead of the satellite. Since Randy Smith, VE3SAT, will be away on extended leave during the first part of the Phase III-A mission, John Fox, W0LER and Ron Dunbar, W0PN, together will be prime command and telemetry stations for Phase III. Randy will join in upon his return. Located approximately 200 miles apart, Ron and John will have a truly redundant system. Each ground station can back up the other. In fact, each can control the other's station from his own location via telephone or radio link. The Minnesota team expects to complete their ground station check-out by late 1978 and begin a one-year training period in preparation for the launch. Included will be several simulations of the launch day.

In order to facilitate writing software for the new spacecraft, Karl, DJ4ZC, has developed a high-level language for the RCA CDP-1802 processor and for the 8080's which will be used at the ground stations. This language, known as IPS (for a German acronym) is a structural language which has some similarities to a language known as F0RTH. Among its features, many of them unique, is the fact that the mnemonics themselves are bilingual. They may be changed from German to English (or vice versa) under software control. AMSAT hopes to make this language available to its members as part of a Phase III package which will also include printed-circuit boards needed for demodulating the engineering beacon telemetry and interfacing to a microprocessor, as well as needed documentation. We still have a way to go before this can be done.

Ron, W0PN, recently went to Germany (taking along and bringing back an 8080 system) to visit Karl and complete needed documentation to interface IPS to the 8080 system he and John, W0LER, are using. This visit also serves to bring Ron up to speed on all the engineering details on the Phase III spacecraft.

SPACECRAFT LABORATORY

1978 NASA and AMSAT signed a contract to jointly that will allow AMSAT to demonstrate to the public uild spacecraft, specifically Phase III. Under the ract, AMSAT and NASA will jointly fund a facility at ce Flight Center to be used by AMSAT.

Chapter 4
Setting Up Your OSCAR Earth Station

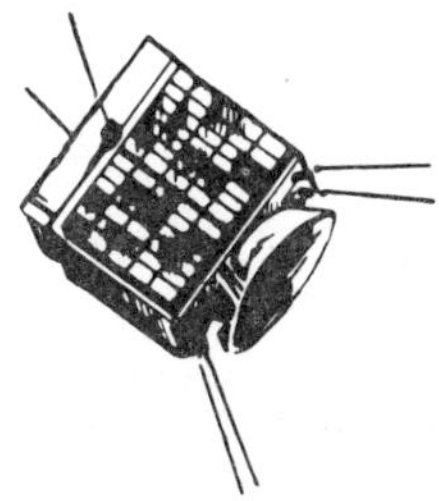

Once you've decided to join the fun of satellite communications, your prime consideration will be securing and arranging a station for operation via OSCAR. Naturally, this consideration will be directly influenced by the particular mode of satellite operation contemplated. Mode A, for example, will merely require a 2 meter transmitter capable of SSB and/or CW operation. Mode J will require the additional capability of receiving signals on 432 MHz.

SELECTING GEAR

Several options are open to the satellite enthusiast at this point. You may purchase transverters for 2 meters and/or 432 MHz; you may homebrew comparable transverters; you may modify surplus high-band transceivers originally used for public service work like police, taxi's, etc.; or you may purchase one or two of the popular all mode VHF transceivers to use with your existing gear. Since the satellites operate duplex fashion on two separate amateur bands, the most effective amateur setups are those arranged with this in mind. Thus, ground stations using two transceivers or a separate transmitter and receiver are situated much better than stations using a transverter and a transceiver. The ability of hearing one's own satellite-returned signals is an extremely important consideration for successful operations.

Since the high frequency satellite gear is often used independent of low-band equipment, it may be located at any convenient place you desire. Several amateurs, for example, have experienced a refreshing change of pace when placing satellite stations in the attic while their low band setups are situated in the den or basement.

ASSEMBLING THE STATION

After the necessary gear is obtained, the satellite enthusiast can begin setting up his earth station. Since the transmitter and receiver frequency dials will be used almost continously, they should be situated within a comfortable reach by each hand. Right handed operators will find operations quite convenient if the transmitter is placed on the left side of the receiver, and the key is placed on the receiver's right side. This eliminates "crossed arms" during passes (Fig. 4-1). Likewise, if the antenna rotor controls and station clock are situated on the left side of the station desk, their use will feel natural and convenient. Since power supplies and broadband linear amplifiers seldom require attention during a satellite pass, they can be placed at any convenient point which doesn't require long cable runs. Frequency relation charts and antenna tracking notes are most effectively used, also, when placed under a glass sheet covering the operating desk.

ANTENNAS

Because satellite operations take place on two amateur bands, two separate antennas are required. Naturally, these "skyhooks" should be designed to produce maximum efficiency for the particular bands used. Also, the antennas should be physically separated by at least 15 feet (mode A) or 4 feet (mode J or B) to prevent any crosstalk or desensitation between the two OSCAR rigs. If mode A operation is contemplated, the 10 meter *downlink* antenna may consist of crossed dipoles or turnstile units. Tri-band beams or loop antennas also perform very well for receive capability.

The criteria for a suitable 2 meter *uplink* antenna is slightly more complex. If a low power transmitter is used (approximately 10 watts), a relatively high gain antenna is usually required. Such antennas will have a narrow beamwidth and require precise azimuth and elevation tracking of the satellite. Antennas in this class are arrays with eight or more elements in either horizontal or vertical planes. If a high power transmitter is used (approximately 150 watts), a smaller and lower gain antenna may be used. These antennas are easily erected or packed for weekend jaunts to the mountains. Additionally, the smaller antennas have a much wider beamwidth which greatly simplifies satellite tracking procedures. Quite often, small antennas perform very well when tilted skyward approximately thirty degrees and rotated in azimuth only (Fig. 4-2).

Fig. 4-1. While this station has a pleasing appeal, it is rather uncomfortable for satellite operation. The 2 meter transmitter (to the right of wattmeter) requires either "crossed arms" for frequency spotting or the downlink receiver (TR-4CW) must be tuned by the left hand.

If feasible, use high power (50 to 150 watts) and low or unity gain transmitting antennas to simplify or eliminate rotor requirements. This more expensive means of operating OSCAR is the most enjoyable method. If you use low power (approximately 10 to 20 watts), don't waste time and money with small antennas. Attempt to acquire an array which will increase your effective radiated power to 100 watts. If, in any case, your signal into OSCAR is stronger than other signals, remove some director elements from the transmitting antenna.

When erecting the satellite antenna, try to select a location where the antenna will not be obstructed. This consideration will govern the approximate height of your OSCAR antenna. If you have a horizon to horizon view in all directions, the antenna can be placed anywhere that's convenient. If you must be selective of favored and lost directions, consider the areas affording the most profitable satellite pass times. Usually this arrangement dicates near-horizon views toward the Northwest, North and Northeast, because most northern hemisphere stations lose two to five minutes of operating time when the satellite crosses the equator. Clear northern skies will be the keynote for successful Phase III operations also.

OSCAR TRACKING AIDS

While many satellite enthusiasts begin their activities using simple equatorial time-crossing data and basic OSCAR Locators, these operators soon realize the advantage of more precise tracking aids. Once these "personalized aids" are incorporated in your OSCAR setup, operations are much smoother and successful.

One inexpensive and highly efficient OSCAR aid is the *satellabe* multi-chart designed by Kaz Deskur, K2ZRO, and distributed by the Ham Radio Publishing group. This satellite plotting device can be used for tracking OSCAR 7, 8, RS, etc., and it provides all necessary information at a glance. Once the Satellabe is calibrated by a daily reference orbit, its concentric scales will indicate precise times of consequent orbits and their positions while the affixed overlay shows the satellite's exact path. This clear overlay is also calibrated in two minute increments.

Satellite operators desiring to go first class can build a visual satellite plotter and add it to their world globe. An elaborate mechanical tracking aid of this nature is described in *Advanced Communications Techniques* available from the American Radio Relay League.

Another form of real time visual tracker which may be added to the station's wall map uses 16 equally spaced Light Emitting Diodes mounted on a framework or overlay placed against the station's map. The LED's are sequentially scanned by a simple 16-bit counter consisting of a low frequency NE555 timer circuit and SN7676 16-bit decoder chip. This arrangement is intended to produce a large, real time equivalent of the popular OSCARlocator, so the framework should be curved to correspond with a particular satellite's path. The LED's are equally spaced from the equator to north pole on this overlay, and their associated wires are connected to proper outputs of the decoder. Once the time base for this "plotter" is properly set (16 counts in 25.75 minutes for equator to pole tracking of OSCAR 8, for example) the LED's will show the satellite's exact location during a pass. This visual indication is an outstanding asset for judging rotor positions, satellite access time, etc.

Numerous other satellite tracking aids are possible. A ceiling, for example, can be covered with a mural of the night sky and the LED's placed across this area to indicate a satellite's path. However, we will leave their design and construction to your imagination. One word of warning: several would-be satellite buffs have become highly involved with calculating orbital information and they never return to the fun world of OSCAR operations. Don't get sidetracked and miss the fun of *operating* OSCAR.

TAPE RECORDING OSCAR

If you have one or two tape recorders which can be pressed into service during OSCAR passes, their use is highly recommended.

Fig. 4-2. OSCAR setup of Paul McClain, N4ZS, features a KLM all mode 2 meter transceiver with accompanying KLM amplifier (right side of desk) for uplinking, and Collins "S" line for downlinking. Satellite antenna is 5-element twist type.

The ability to relive a previous satellite pass is both enjoyable and beneficial in acquiring additional satellite knowledge. If one recorder is available, place its mike at a point which permits monitoring of transmitted and received signals, and allow the unit to record continuously during a pass. One never knows exactly what will occur during satellite activities, and this "extended operation" aspect assures such phenomena will be recorded. Likewise, when reviewing a satellite pass, one may realize a previous oversight which restricted operations.

If special modes of operation like SSTV or RTTY are contemplated, two tape recorder setups will definitely be useful. Prerecorded programs may be cued to their exact beginning and placed in one recorder for instantaneous transmissions while the other recorder monitors all signals appearing at the downlink receiver's output. After a pass has concluded, the tapes may be rewound and evaluated. The "automatic operation" capability afforded by tape recorders also adds a truly professional and enjoyable dimension to satellite operations.

BYPASSING COSTLY POWER SUPPLIES

Medium and high power amplifiers require large, relatively expensive power supplies which often present obstacles to the amateur with a limited budget. Most of these amplifiers are transistorized units which require 12 volt, high current power sources. An alternate solution to their problem involves using an old motorcycle or auto battery to power the amplifier. The battery can be trickle charged with a simple homebuilt mini-charger during periods of non use. Thus, most "junk" batteries will produce excellent results when used in this manner.

Another possibility which is particularly appealing to the clean car buff or home mechanic-type amateur is temporarily borrowing his auto's battery for use during satellite passes. If the car's engine compartment is relatively clean and unconfined, the fully-charged battery may be removed or reinstalled in less than 10 minutes. Four cylinder cars are especially convenient for these tactics. Don't try operating more than a couple of satellite passes before reinstalling the battery and recharging it, or you may be left stranded the next morning.

Chapter 5
In's and Out's of Satellite Operation

The keynote to getting maximum enjoyment from each OSCAR pass is being prepared and ready for any activities and conditions which may suddenly crop up. Expressed quite simply, it's better to be overly prepared than to be caught with your accessories down. The uplink transmitter's mike and key should be placed ready for instant use, as circumstances dictate. Likewise, if SSTV operation is contemplated for a certain pass, the Slow Scan gear should be switched on and its operation checked *before* the satellite's signals are heard. Tape recorded SSTV and tape "blanks" for receiving use should also be cued to their beginning points. Satellite signals will be weak at first, so your best bet is initially operating CW. Your satellite returned signal will vary in strength and frequency during this time, so a straight key will prove much better than a keyer or semi-automatic "bug" for controlling dot/dash length and for mental monitoring of your fist. If you try to use the satellite returned signal for CW monitoring, U's and V's will sound like 4's, etc.

As the satellite rises in the sky, its translated signals will increase in strength. This will be the prime time for operating SSB or SSTV. Don't waste any of this time, as the satellite will soon pass overhead and progress toward the far horizon. Stay particularly alert, also, to avoid hopelessly pushing your SSB capability when signals begin to fade. Switch back to CW immediately and bring your DX tactics into play until the satellite moves out of range. Then after the pass has concluded, you can relax and review your notes or tapes.

When zeroing or frequency-spotting SSB stations, it's usually helpful to first send V's on CW, and then "talk yourself" onto the exact frequency desired.

Answering CQ's provides much better results than calling CQ. This is because other stations may have difficulty zeroing your frequency and thus call CQ, hoping other satellite operators will answer them.

Try to maintain a running list of various stations and their approximate frequencies during a pass to avoid missing DX when you're elsewhere calling CQ.

If your satellite antenna is somewhat restricted in size or your satellite-returned signal is marginal, try operating during QRP nights (Sunday nights). The results may amaze you.

If you need to gain experience zeroing frequencies, calibrating charts, aiming antennas or calculating doppler shift between AOS and LOS, try brief test transmissions during experimental nights. It is suggested, however, that advance authorization be obtained from the ARRL or AMSAT before using this technique.

A speech compressor (Fig. 5-1) will perform two outstanding purposes when used with the OSCAR uplink transmitter. First, it will provide the additional audio punch often necessary for reliable communications. Secondly its gain control will provide a means of varying SSB uplink output power when the satellite passes your point of maximum strength (the time when SSB operation is usually employed).

Fig. 5-1. High quality speech compressors are quite valuable for OSCAR satellite work, as they allow maximum use of available RF power. The Vomax unit shown here produces up to 10 db increase in "talk power."

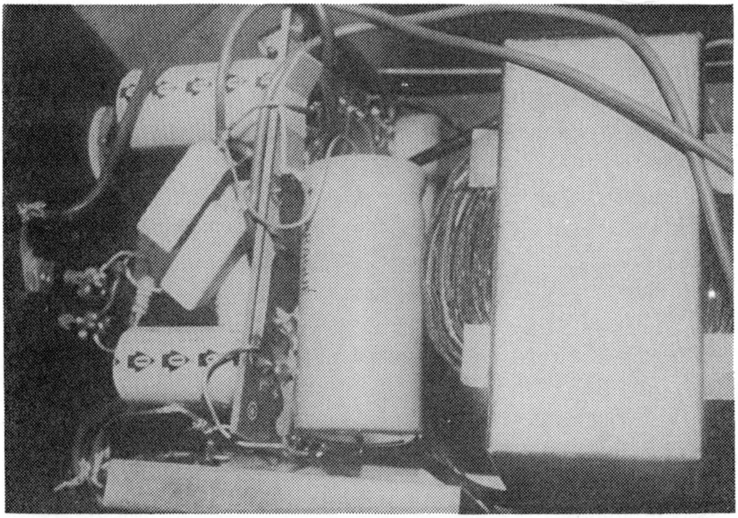

Fig. 5-2. The transformer used in this power supply was quite stout, so an additional regulator and pass transistor were added to double output current capability. This transistor can be seen bolted to rear of power supply. Finally, a large output capacitor was added to further increase current-producing ability. Additional capacitor is between transformer and circuit board. Large sheet at bottom of the picture is original heat sink used in the supply.

If your linear amplifier's power supply is shy on current delivering capability, an extra computer type high capacitance capacitor (30,000 to 70,000 mfd typically) placed across the supply's output terminals will provide slightly more output power for SSB or fast CW operation (Fig. 5-2).

If your financial status places you uplink signal in the QRP category, consider adding db's with antenna elements rather than a higher power transmitter.

DXING THE SATELLITES

Successful DXing via OSCAR 7 or 8 usually requires the ability to concentrate one's maximum ERP at low-horizon positions of satellite. These passes will allow the satellite to be midway between you and the particular DX area you desire to contact. As an example, there is a brief period as OSCAR crosses the Atlantic Ocean when communication capabilities of Europe and the U.S. overlap. A similar situation is also produced between the U.S. and South/Central America as OSCAR crosses the equator. Other DX possibilities for your particular QTH can be found by simulating a low horizon, distant pass on your OSCARLocator or Satellabe and then studying the other areas which may also access the satellite during this time.

Since DX passes will be low on one's horizon, satellite access time will be brief. The exact AOS, access and LOS times will be instantly apparent on your OSCARLocator.

These aspects can be used to definite advantage by the DXer. The brief periods of these passes fall precisely according to orbital calculations, thus producing a high success/operating ratio for the radio amateur. Likewise, satellite propagation is highly predictable and unaffected by sunspot activity. This situation results in 99 per cent accuracy of DX predictability.

A high Effective Radiated Power level and selectable antenna polarization are particularly useful for long distance satellite communications. While right or left hand circular antenna polarization is used for regular satellite communications, the DX operator realizes a theoretical 3db gain by using either vertical or horizontal polarization as the case requires. Although not an absolute fact, it has been found that Faraday rotation of distant satellite signals usually require horizontal polarization of earth-based transmitting antennas on low horizon passes.

Another highly noteworthy consideration for DX operations via satellite is a high gain, low noise preamplifier for downlink signal reception (Fig. 5-3). When a satellite is at a distant point, all downlink signals will be weak; particularly the distance station you wish to contact. Also, your downlink signal *should not* be stronger than other satellite signals. This means your receiving capabilities should be optimum, and a superb preamplifier/antenna is the keynote for this operation.

Use your known and proven frequency relation charts for instant "frequency spotting" during DX passes. The time associated with these operations is usually brief, and you want every possible asset working for you. We're discussing OSCAR 7 and 8 operating at this point; Phase III is a totally different ball game.

Stay aware of special VHF conditions like Tropospheric ducting, sporadic E skip, Meteor showers, etc. and use these aspects to enhance your operations. Many times, these brief openings allow extended distance satellite communications with high degrees of success. Since OSCAR 7 and 8 satellites passes are relatively brief, and much of today's gear is solid state, a quick check of nightly orbits is easily accomplished.

A portable or mobile OSCAR setup can be advantageous for DX operations, particularly when the home QTH is blocked by trees or hills in the desired directions. Realizing that foliage and terrain plays havoc with RF energy at VHF frequencies, many amateurs have carried their gear to hilltop locations or oceansides to enjoy DX

Fig. 5-3. High gain, low noise preamplifiers boost satellite-transmitted signals and make OSCAR operations a true pleasure. Their use of all operations via the satellite is highly recommended. Shown here are inexpensive, yet effective 2 meter and 10 meter preamps.

passes. The results have been quite gratifying. While such operating tactics are quite difficult for low band operations, their use for VHF and OSCAR are very feasible and highly rewarding!

A tilted antenna mast is another beneficial consideration for the OSCAR DXer wishing to bypass the usual two rotor (AZ/EL) setup. Simply tilt the mast from the ground level so the satellite antenna points approximately thirty degrees skyward when orientated toward the passes highpoint. Then, when the antenna is pointed toward either extreme of this direction, the array will automatically assume the required low angle of orientation (Fig. 5-4).

DOUBLE LINKING SATELLITES

The ability to simultaneously operate through two amateur satellites is a unique aspect of satellite communications which can be enjoyed by any properly informed and equipped OSCAR operator. Naturally, the two satellites involved in such *double links* must be in transmission/reception range of your QTH, and in the proper modes to allow the output of one bird to access the other bird. Such conditions occur on the average of once each month.

Once a possible double link situation has been determined, a frequency correlation chart should be drawn and double checked for satellite compatibility. Next, a beam positioning plan should be drawn to assure correct aiming during the passes. Finally, AMSAT/ARRL should be pre-notified of your plans.

Under ideal circumstances both satellites will be in the same general direction, with one being farther from your QTH than the other. If the receiving satellite is farther, your DX transmitting tactics will be necessary. If the transmitting satellite is farther, you'll need a hot receiving preamplifier and high gain antenna. The time of such *interphases* is usually limited by LOS of the most distant satellite, so plan your time accordingly and work fast at the start to insure success. Naturally, you should consider tape recording the full event for future study and evaluation.

I would like to re-emphasize that exact frequency relations between each satellite must be possible for double linking to succeed. Simply assuming the mode J output of one satellite can access the other satellite's mode B input is insufficient. OSCAR 8's mode J output, for example, is on 435 MHz while OSCAR 7's mode B input is on 432 MHz. As with all satellite operations, careful planning significantly reduces the possibilities of errors. Consider every aspect from a negative viewpoint, and your success ratio will be very high.

OPERATING OSCAR MOBILE

Satellite operation is one phase of amateur radio which is perfectly adaptable to mobile or fixed portable operation. Since an efficient satellite station is relatively small in size and doesn't require extremely high power, the gear easily can be placed in a vehicle. This vehicle, which may be a car, boat, etc., then can reach a location which affords perfect horizon to horizon views, and its capabilities will equal that of regular home-base operations. If the mobile satellite operator has a choice of gear to use, he will realize optimum results with 100 to 200 watts RF levels, high gain receiving preamplifiers and omni-directional antennas.

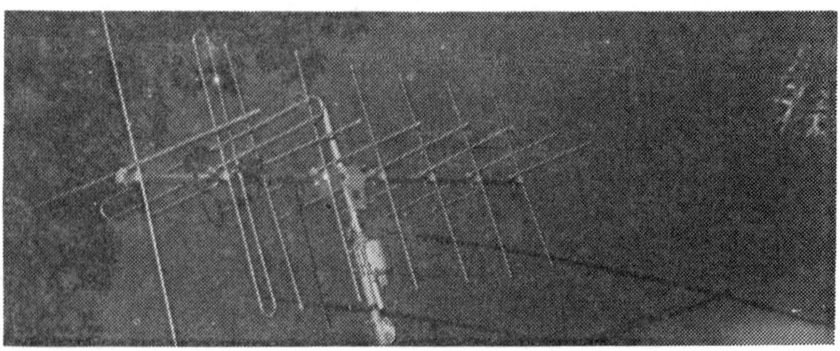

Fig. 5-4. Tilted antenna mast and Spectrum International twist antenna provide single rotor tracking technique for this OSCAR setup. Triband beam in background is used for mode A downlink reception.

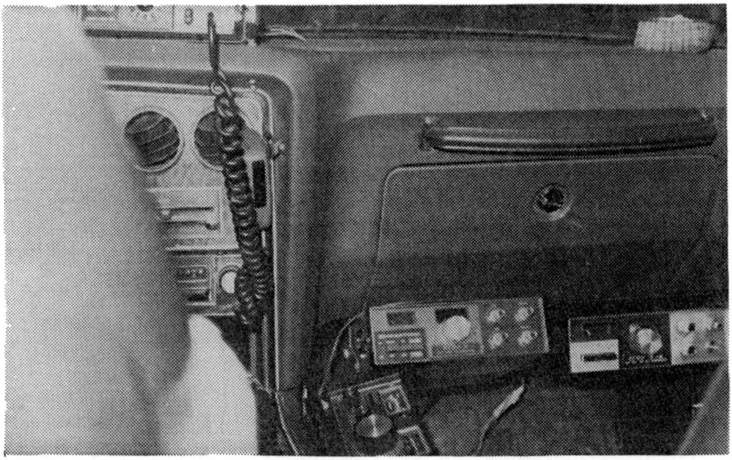

Fig. 5-5. Lavish OSCAR mobile setup of Fred Merry, W2GN. Equipment includes KLM Echo II 144 MHz SSB/CW transceiver, KLM Echo 70 432 MHz SSB/CW transceiver, Atlas 210X 3.5 to 30 MHz transceiver, ICOM 20 146 MHz FM transceiver, plus trunk mounted amplifiers for 144 MHz and 432 MHz. Antennas are ⅝ths wave whips on 144 MHz and 432 MHz and Hustler whips for 80 through 10 meter bands.

Alternately, a low power setup in conjunction with a beam antenna for uplink signals can be used. The beam can be placed on a simple knock-down tripod and hand rotated during a pass. A beam antenna and hand tracking methods are quite feasible for mobile OSCAR operations because the auto is usually stopped during such activities. It's simply too difficult to track doppler shift, zero frequencies, and follow time relationships while moving down a busy road. Fixed mobile operations from hilltops, however, are an absolute blast of enjoyment (Fig. 5-5).

SSTV VIA OSCAR

Slow Scan TV operators can combine the best of two exciting frontiers when operating video via satellite. These operations require a working knowledge of both modes, but the overall results are well worth their efforts. This situation will particularly hold true with the advent of our Phase III satellites, as video communications with many countries on a reliable basis will be possible.

SSTV and RTTY operators are aware of the 100 per cent duty cycle associated with their modes of communication, and they should always reduce power at least 60 per cent when operating via satellite. Additionally, the operator should constantly monitor his downlink signal to assure that it is not stronger than other translated signals.

The most reliable and enjoyable means of operating SSTV via OSCAR is by pre-recording specific programs or pictures before a pass, then transmitting these tapes as desired. A second tape recorder is highly advisable for recording information returned via satellite without the fumbles of "one reader" setups. Naturally, the SSTV equipment should be switched on and its operation checked before a pass begins. Then, the operator should watch the satellite's activities immediately following AOS to assure that fast fades etc. will not disrupt video communication. As the satellite moves high in the sky, the time will be optimum for SSTV operations. These transmissions should be limited to 16 or 24 seconds, with frequent breaks to insure the other party is copying the information. Video only transmissions or SSTV CQ's via OSCAR are discouraged, as they tend to misuse satellite power and spectrum. The desired technique involves initially contacting other SSTVers via SSB, then switching to Slow Scan for the brief video transmissions. Notifying the ARRL and/or AMSAT of your SSTV activity via satellite is also considered good operating practice.

A good downlink receiver and antenna are particularly advantageous for SSTV operations, as consistent copy for several seconds is an absolute must. The SSTV operator also will realize best results when transmitting white images on black backgrounds. This method has a tendency to reduce interference caused by deep fades or high noise levels.

THOUGHTS ON PHASE III

Our new group of satellites will result in a totally new ball game for all amateurs in the northern hemisphere. The possibility for communications between various countries may produce some interference problems thus some "special considerations" should prove fruitful. Keeping in mind that European countries are 7 to 9 hours ahead of our time while Far East countries are 7 to 9 hours behind our time, the enterprising amateur should consider varying his sleeping and operating times to avoid stateside competition and interference. Another possibility involves split-sleeping, and checking satellite activity at various and apparently inopportune times. Since most satellite gear is solid state design, its instant-on capability is very helpful for these brief activity checks. There's also a good possibility that most operators will ignore the Phase III satellite during its brief travel around the southern hemisphere, leaving such times wide open for QSO's and test by alert satellite users. Don't miss any of the action!

Chapter 6 Satellite Tracking Simplified

Our amateur satellites can serve no useful purpose if we do not know precisely when and where to find them. Several variations of this basic information are available to the satellite operator, thus he may select from a lavish array of data sources. Skip Reddman, W6PAJ, produces an elaborate yearly calandar which lists all orbital information for each day's activities of a specific satellite. This data is pinpoint correct, and it makes OSCAR operations a truly enjoyable affair. Contact AMSAT at P.O. Box 27, Washington, D.C. 20044 for details on how to get a calendar.

The first orbit of each day's operation of OSCAR 7 and 8, known as reference orbits, are published in *QST, 73,* Worldradio and in AMSAT's quarterly magazine. The satellite enthusiast may calibrate his Satellabe, OSCARlocator, or homebrew tracking aid by this information and obtain pertinent data on passes for that day.

Finally, the satellite enthusiast may engage the resources of a computer hobbyist and obtain his own computer program of satellite crossings, personalized for his QTH.

ESTABLISH AN OPERATING PLAN FOR PASSES

Once the equatorial crossing times for specific satellite passes have been secured, the satellite buff may take an additional step and plan each operation to be conducted during a pass. This information may consist of projected equinox time, beam position adjustments for each two minute periods, approximate signal strengths expected during these times, planned periods of SSB or SSTV activities, time predictions for loss of satellite returned signals, etc. Then, during that particular satellite pass, the predetermined data is placed at a conspicuous point in the station, along with the frequency relation charts to be used, and used to coordinate all activities (Table 6-1).

Often, out-of-doors operations and/or public demonstrations are conducted during a satellite pass. An active OSCAR operating plan is useful during these times. One amateur, armed with antenna tracking data, can handle this function while one or two equipment operators provide operating/mode schedules and frequency relations.

CALCULATOR PROGRAMS FOR TRACKING OSCAR

While the satellite orbital calculations and tracking aids already mentioned are quite sufficient for OSCAR operations, many amateurs want to use their electronic calculators. With this in mind here are some amateur-developed satellite calculation methods which can run on pocket calculators. Since all the work associated with these programs has been performed, you can merely load your

Table 6-1. Sample Worksheet for OSCAR Pass.

January 22 '79 OSCAR 8 Orbit #4497 Mode A				
0215 GMT	76°	EQX	Turn on Preamp	
0217	SSE	AOS	Weak signs.	29.432
0219	SSE	20° EL.	CW DW	to
0221	ESE	30° EL	CW DX	29.452
0223	ESE	60° EL	Monitor band	
0225	Satellite overhead		Brief loss of signs	
0227	N	50°	SSTV/SSB 29.462	watch
			to	power
0229	N	20°	SSTV/SSB 29.472	level
0231	NW	Follow satellite out	CW DX	29.432
				to
0234	LOS		CW DX	29.452

Table 6-2. Calculator Instructions for WØSL Circular Orbit Program.

STEP	INSTRUCTIONS	INPUT DATA/UNITS	CALCULATOR KEYS		OUTPUT DATA/UNITS
1	Load Program (use 2 sides of HP Cards)				
2	Store Orbit parameters	STA. Long	STO	0	DEG.
	A. Key in manually	360/P	STO	3	-
	or	inclination	STO	4	DEG.
	B. Load in from Data Card	STA Lat.	STO	5	DEG.
		R/(r +h)	STO	7	-
3	Enter Longitude of North-bound Node	NBN long.	D		DEG.
4	Enter minutes after NBN (+ or –)	T	E		MINS.
	OR				
	Enter NBN Clock Time	HH.MMSS	ENT		MINS.
	Enter current (start) time	HH.MMSS	C		MINS.
5	Compute AZ-EL and Display in Real Time at one minute intervals		A		AZ-EL
	OR				
	Compute Latitude and longitude of Subsatellite Point (SSP) and AZ-EL. A Negative Latitude is south and a negative Longitude is east. Display		B		SSP LAT. SSP LONG. AZ-EL
	in Batch mode. Press R/S key to update Batch note in 2 minute increments.		R/S		MINS.
6	To input new data, Press R/S key to stop run and go to Step 3. If no change is required for Step 3, go to Step 4. All inputs may be corrected immediately by re-executing the appropriate step. Steps 3 and 4 may be reversed in sequence.		R/S		

calculator directly with information from the following pages and you're set.

The WØSL AZ/EL Tracking Method

The following calculator programs were produced by Roy D. Welch, WØSL, and they provide a continuous display of antenna tracking data which can be used for any pass of a specific amateur satellite. Using this information, Roy has acquired Worked All States certificate #22, and contacted over 40 countries via OSCAR. Since the *circular orbits program* can be used with satellites using inclination angles of zero to 180 degrees, it is applicable for OSCAR 7, 8, the U.S.S.R. satellite RS, or any other circular orbit craft (Table 6-2). The *elliptical orbits program* has been extensively tested on paper, and appears to be very accurate. See Tables 6-6 through 6-9. Both WØSL programs will work on the Hewlitt-Packard HP67 or HP97 calculator. These programmable calculators are quite popular among high school and college students and are available in department stores.

The program will run in either of two selectable modes; Real Time and Batch modes (Table 6-3). When run in Batch mode, the program will display azimuth and elevation of ground station coordinates to an orbiting satellite such as OSCAR 7 or OSCAR 8. The display will remain stable until the **R/S Key** is depressed. The program will then select a time two minutes later and display new coordinates (Table 6-4). The Batch mode is for use when you wish to make listings of AZ/EL information for various equatorial crossings.

The Real Time mode allows you to enter the longitude of the Northbound (Ascending) Note (NBN), NBN clock time and current clock times. The program will then begin to display current AZ/EL coordinates and update them for new coordinates at one minute intervals. Calculation takes about 21 seconds. The remaining 39 seconds of each minute is consumed by the display output (looping). This mode is useful in that it eliminates the pre-orbit AZ/EL pencil labor. Either mode may use the clock time input or the simple minutes after NBN approach to set the time factor. See Table 6-5 for a sample problem.

The circular orbit program calculates azimuth and elevation angles (AX/EL) for tracking circular orbit crafts like OSCAR 8, etc. The necessary satellite orbital data and ground station coordinates may be input manually or via data programming cards. The resultant calculator display gives tracking data for each approximate two minutes of a pass.

Given: 1) Radius of earth, R (arbitrary units)
2) Satellite orbital parameters
   (a) Inclination angle in degrees, Incl.
   (b) Altitude (same units as R), H
   (c) Period (minutes), P
   (d) Longitude of Northbound Node, NBN V_x (0 to 360 degrees West).

Table 6-3. WØSL Circular Orbit Program.

(continued on pages 74-76)

STEP	KEY ENTRY	KEY CODE	NOTES
001	F LBLA	31 25 11	Real Time Start
	H CFO	35 61 00	
	G + O O	22 00	
	* F LBL B	31 25 12	Batch Start
	H SFO	35 51 00	
	G to 0	22 00	
	* F LBLC	31 25 13	Enter clock times and calculate
	H X⇄Y	35 52	Time difference in hours following NBN time.
	CHS	42	
010	H h.ms +	35 83	
	F H	31 74	
	6	06	Change to minutes
	0	00	
	X	71	
	•	83	Add 24 seconds to allow
	4	04	for run time before
	+	61	display.
	* F LB LE	31 25 15	Enter minutes after NBN
	STO 1	33 01	time.
020	H RTN	35 22	
	* F LBL D	31 25 14	Enter NBN longitude
	DSP1	23 01	
	STO 2	33 02	= NBN longitude
	H RTN	35 22	
	* F LBLO	31 25 00	
	DSPO	23 00	
	H SFI	35 51 01	
	RCL 1	34 01	T
	RCL 3	34 03	360/P
030	X	71	
	STO 6	33 06	360T/P
	F SIN	31 62	
	RCL 4	34 04	Incl (inclination)
	9	09	
	0	00	
	G X≤Y	32 71	Is incliantion 90 degrees?
	H SF2	35 51 02	
	H R↓	35 53	
	F SIN	31 62	
040	X	71	
	G SIN −1	32 62	
	STO 8	33 08	= Op
	STO D	33 14	= Op (SSP Latitude)
	H CF3	35 61 03	
	F X<O	31 71	Is SSP Latitude negative?
	H SF3	35 51 03	
	RCL 6	34 06	360T/P
	F COS	31 63	
	RCL 8	34 08	Op

Table 6-3. WØSL Circular Orbit Program
(continued from page 73 and on page 75).

STEP	KEY ENTRY	KEY CODE	NOTES
050	R COS	31 63	
	F ≠ O	31 61	Is divisor = 0?
	÷	81	
	F GSB 3	31 22 03	
	G COS −1	32 63	
	H F?2	35 71 02	Was inclination > 90 degrees?
	H F?2	35 71 02	
	H F?3	35 71 03	Was SSP Latitude negative?
	H F?3	35 71 03	
	CHS	42	
060	RCL 1	34 01	
	4	04	
	÷	81	
	+	61	
	RCL 2	34 02	Vx
	+	61	
	STO E	33 15	= Vp (SSP Longitude)
	RCL Ø	34 00	Vq
	−	51	
	STO 9	33 09	= V
070	F SIN	31 62	
	O	00	
	X ≤ Y	32 71	is SIN V < Ø?
	H CF1	35 61 01	
	RCL 9	34 09	V
	F COS	31 63	
	RCL 8	34 08	Op
	F COS	31 63	
	X	71	
	RCL 5	34 05	Oq
080	F COS	31 63	
	X	71	
	RCL 8	34 08	Op
	F SIN	31 62	
	X	71	
	+	61	
	STO 9	33 09	= COSc
	G COS^{-1}	32 63	
090	F SIN	31 62	
	STO 6	33 06	= SINc
	RCL 8	34 08	Op
	F SIN	31 62	
	RCL 9	34 09	COSc
	X	71	
	−	51	
	RCL 5	34 05	Oq
100	F COS	31 63	
	RCL 6	34 06	SINc
	X	71	
	F X ≠ O	31 61	Is divisor = Ø?
	F GSB 3	31 22 03	

Table 6-3. WØSL Circular Orbit Program
(continued from page 74 and on page 76)

STEP	KEY ENTRY	KEY CODE	NOTES
	G COS^{-1}	32 63	B
	H F?1	35 71 01	
	GTO 1	22 01	
	CHS	42	
110	3	03	
	6	06	
	0	00	
	+	61	
	GTO 2	22 02	
	* F LBL 3	31 25 03	Subroutine to protect
	1	01	against error half due
	G X ≤ Y	32 71	to rounding producing
	H RTN	35 22	value slightly greater
	CHS	42	than 1 or less than
120	G X ≤ Y	32 71	−1 or COS^{-2} command.
	H X⇄Y	35 52	
	H RTN	35 22	
	* F LBL 1	31 25 01	Program timing equalization
	H CF3	35 61 03	steps
	H CF3	35 61 03	
	H CF3	35 61 03	
	H CF3	35 61 03	
	H CF3	35 61 03	
	H CF3	35 61 03	
130	H CF3	35 61 03	
	H CF3	35 61 03	
	H CF3	35 61 03	
	* F LBL2	31 25 02	
	F RDN	31 24	
	STO 8	33 08	= AZ
	RCL 9	34 09	COSc
	RCL 7	34 07	R/(R + h)
	−	51	
	RCL 6	34 06	SINc
140	G→P	32 72	Equivalent to TAN^{-1}
	H R↓	35 53	= EL
	EEX	43	Convert EL to EL
	2	02	
	÷	81	
	DSP 2	23 02	
	F RDN	31 24	
	F X > O	31 81	Is EL positive?
	GTO 4	22 04	
	CHS	42	Change sign of EL
150	H SF2	35 51 02	If negative
	F LBL4	31 25 04	
	RCL 8	34 08	AZ
	+	61	= AZ .EL
	H F?2	35 71 02	If EL was negative, change
	CHS	42	sign of AZ.EL
	STO 8	33 08	= AZ.EL for display

Table 6-3. WØSL Circular Orbit Program

(continued from page 75).

STEP	KEY ENTRY		KEY CODE	NOTES
	H F?0	35	71 00	Is this a batch run?
	GTO 6		22 06	
	H PAUSE		35 72	Real Time AZ EL display
160	H PAUSE		35 72	
	H PAUSE		35 72	
	H PAUSE		35 72	
	H PAUSE		35 72	
	H PAUSE		35 72	
	H PAUSE		35 72	
	H PAUSE		35 72	
	H PAUSE		35 72	
	H PAUSE		35 72	
	H PAUSE		35 72	
170	H PAUSE		35 72	**Note:** HP67 users may re-
	H PAUSE		35 72	place steps]59-] 84 with
	H PAUSE		35 72	F-X- (3184) commands if
	H PAUSE		35 72	desired. If this option
	H PAUSE		35 72	is chosen, the value in
	H PAUSE		35 72	step 185 should be changed
	H PAUSE		35 72	to 2.
	H PAUSE		35 72	
	H PAUSE		35 72	
	H PAUSE		35 72	
180	H PAUSE		35 72	
	H PAUSE		35 72	
	H PAUSE		35 72	
	H PAUSE		35 72	
	H PAUSE		35 72	
	4		04	Timing loop adjust
	H STI		35 33	
	* F L8L5	31	25 05	Timing Loop
	F DSZ		31 33	
	GTO 5		22 05	
190	1		01	Increment T by one minute
	STO +1	33	61 01	
	GTO 0		22 00	Start new calculation
	* FLBL6	31	25 06	Set up batch display
	DSP1		23 01	
	RCL D		34 14	SSP Latitude
	F-X-		31 84	Display SSP Latitude
	RCL E		34 15	SSP Longitude
	* F LBL7	31	25 07	
	H		35 73	
200	F D←		31 73	
	G X Y		32 81	Is Long. > 180 degrees?
	GTO 8		22 08	If so, subtract 360 degrees.
	2		02	
	X		71	
			51	
	GTO 7		22 07	Go check again
	* F LBL8	31	25 08	
	H R		35 53	= Corrected SSP Long.
	STO E		33 15	Display SSP long.
210	F-X-		31 84	AZ.EL
	RCL 8		34 08	
	DSP 2		23 02	Display Batch AZ.EL
	R/S		84	Increment T by 2 mins.
	2		02	
	STP +1	33	61 01	
	RCL 1		34 01	Display new T
	F-X-		31 84	
	GTO Ø		22 00	
	H RTN		35 22	Note: steps 140 and 141 will
220				process EL = ±90 degrees with-
				out halt (error) produced by
				÷ TAN [1]

Table 6-4. Calculator Register Data for WØSL Circular Orbit Program.

0 = STA long Vq (degrees)
1 = minutes after T (minutes)
2 = NBN longitude Vx (degrees)
3 = 360/P
4 = inclination Incl. (degrees)
5 = STA latitude Oq (degrees)
6 = (used)
7 = R/(r + h)
8 = (used)
9 = (used)
A, B, C, (open). D, E, I (used)

Calculator labels for previous program

A = Real Time Start
B = Batch Start
C = Enter Clock Times
D = Enter NBN Longitude
E = Enter minutes after NBN
Lables Ø through 9 are used by program

Additional calculator notes

Trig. function set at degrees
Display Function set at fixed :n2
Flags Ø through 3 set at off : Batch run flags as follows:
Ø = Batch run, 1 = SIN V ≤ Ø, 2 = used, 3 = used.

3) Earth ground station coordinates.
   (a) Latitude (0 to 90 degrees. + North – South). O_q
   (b) Longitude (0 to 360 degrees West). V_q

4) Time before or after Northbound note (NBN). T (± minutes), or NBN and current clock times.

Program Computes:

1) Relative earth coordinates of subsatellite point (SSP) at time T.
   Op = SSP Lat., Vp = SSP Long.
2) Spherical angles between ground station and SSP.
3) Azimuth and elevation coordinates.

Equations:

1) $Op = SIN^{-1}\left[SIN\ INCL\ SIN\ (360\ T/P)\right]$
2) $Vp - Vx + COS^{-1}\left[COS(360\ T/P)/COS\ Op\right] + T/4$
3) $V = Vp - Vq$
4) $C = COS^{-1}\left[SIN\ Op\ SIN\ Oq + COS\ Op\ COS\ Oq\ COS\ R\right]$
5) $B = COS^{-1}\left[SIN\ Op - SIN\ Oq\ COSc)/(COS\ Oq\ SINc)\right]$
6)a) $AZ = \text{if } SIN\ V < Ø$
6)b) $AZ = 360 - B \text{ if } SIN\ V \geq Ø$
7) $EL = TAN^{-1}\left[(COSc - R/Rth)/SINc\right]$

Special Notes:

1) Minutes after NBN, T can be entered as negative value, i.e., before NBN, but not before T = −P.
2) Clock times must be from same day, i.e., not separated by midnight, unless times after midnight are stated as 24.0105, 24,1520, etc.

Table 6-5. Sample Problem for Use with Circular Orbit Program.

Input data (OSCAR 7 parameters used for example)
R = 6356.914 Km
Incl.=101.7
H = 1460 Km
P = 114.945161 minutes
Oq = 38.61260331 degrees North
Vq = 90.51447368 degrees West
STO Ø = Vq = 90.51447368
STO 3 = 360/P = 3.131928276
STO 4 = Incl 101.7
STO 5 = Oq = 28.61260331
STO 7 = R/(r + h) = .813225478

Real Time Example

OSCAR 7 Orbit 14286 December 20, 1977. NBN at 0045.09 GMT.
Vx = 66.9 degrees West.
66.9 [D] → − 66.9 NBN longitude, Vx
00.4509 [Enter↑] NBN clocktime (HH.MMSS)
00.4400 [C] → − 0.8 current (start) clocktime (HH.MMSS)
[A] → −146.10******AZ = 146°, EL = −10)°
→ −145.07******AZ = 145°, EL = −7°
→ −144.05******AZ = 144°, EL = −5°
→ −143.02******AZ = 143°, EL = −2°
(etc) calculated and displayed at one minute intervals in Real Time.

Batch Mode Example

66.9 [D] → − 66.9 NBN Longitude, Vx
[CHS] [E] → −1.0 minutes after NBN = 1 (i.e. 1 minute before NBN)
[B] → −3.1***SSP Latitude = −2.1° (i.e. 3.1 degrees South)
66.0***SSP Longitude = 66.0 (i.e. 66 degrees west)
−146.11 AZ = 146°, EL = −11° (Below horizon)
[R/S] 1.00*** Minutes after NBN = 1
3.1 *** SSP Latitude = - 3.1°
67.8***SSP Longitude = 67.8°
−144.05 AZ = 144°, EL = −5°
[R/S] → − 3.00*** Minutes after NBN = 3
9.2 *** SSP Latitude =9.2°
69.6***SSP Longitude = 69.6°
142.01 AZ = 142°, EL = 1° (above horizon)

Table 6-6. Calculator Instructions for WØSL Elliptical Orbit Program.

STEP	INSTRUCTIONS	INPUT DATA/UNITS	CALCULATOR KEYS	OUTPUT DATA/UNITS
1	Load Program (2 sides of card)			
2	Store Orbit parameters A. Key in manually or B. Load from data card	J/minutes P/minutes I rad. LQ/rad VQ/rad a/Km p/Km E	STO 0 STO 1 STO 2 STO 3 STO 4 STO 6 STO 8 STO E	
3	Enter longitude of NBN	Vx/Degrees	D	
4	Enter clock time of NBN	T/HH.MMSS	C	
5	Enter selected true anomaly (may range from 0-360 degrees)	Ø/degrees	E	
6	Computer clocktime and AZ-EL coordinates at true anomaly =Ø. Display clocktime. Hold this point until actual time agrees with displayed time before going to Step 7 if real time AZ-EL coordinates are desired. Otherwise proceed with Step 7.		A	HH.MMSS
7	Display AZ-EL coordinates at true anomaly = 0.		R/S	AZ.EL
8	Depress R/S key to increment true anomaly by ten degrees and run new calculations.		R/S	New Ø
	Note: Step 3, 4, and 5 may be entered in any sequence or corrected at any time.			

3) Internal Clocking times may vary slightly between various calculators. This may necessitate altering the timing delays in the Real Time mode. The addition or deletion of a "pause" command in the Real Time AZ/EL display steps will add or delete about one second. Value in timing loop adjustment step may be increased or decreased (do not use zero) to fine tune timing delay loop.

Table 6-7. WØSL Elliptical Orbit Program (continued on pages 81-85).

This program calculates azimuth and elevation angles (AZ/EL) for tracking Phase III satellites in elliptical orbit. Orbital parameters and ground station coordinates are input manually or via data card. Program computes AZ/EL and clock times for incremental true anomalies.

Given: 1) Radius of earth, R (Km)
2) Satellite orbital parameters
(A) Inclination angle, I (degrees)
(B) Altitude of basic orbit, hp (Km)
(C) Period, P (minutes)
(D) True anomaly, Ø (degrees)
(E) Longitude of Northbound Node (NBN), Vx (degrees)
(F) Time from perigee to NBN, J (minutes)
3) Earth ground station coordinates
(A) Latitude, Lq (degrees North +, South −)
(B) Longitude, Vq (degrees west)
4) Clock time of NBN, T(HH.MMSS)

Program computes: 1) Relative coordinates of subsatellite point (SSP) for true anomaly, Ø.
2) Spherical angles between ground station and SSP.
3) AZ-EL coordinates and clock time for each true anomaly value, Ø incremented in ten degree increments.

Equations:
1) A ≈ (P² .36 4085.10⁶) 1 3 Km (Big half axis)
2) E = 1- (hp + R)/a (eccentricity)
3) p = A (1 − E^2) Km (parameter)
4) r = p/(E COS 0 + 1) radius vector at true anomaly = Ø)
5a) E = COS $^{-1}$ [(1 − r/a)/E] (Eccentric Anomaly if Ø <π)
5b) E = 2 − COS $^{-1}$ [(1 − r/a)/E] (Eccentric Anomaly if Ø ≥ =)
6) T = (E SINe) P/2 π minutes (time after perigee)
**7) h = r − R Km (altitude)
**8) X = (r^2 − R^2) 1/2 Km (maximum altitude)
**9) s = TAN $^{-1}$ (X/R) (range of maximum range)
10) Lp = SIN −^{1} [SIN I(− COS Ø)] (I = orbit inclination)
11a) Vp = Vx + (T−J) π/720 + COS $^{-1}$ (SINØ/COS Lp) (*See Note 1)
11b) Vp = Vx + (T − J) π/720 + COS $^{-1}$ (SIN Ø/COS Lp) + π (*See Note 2)
12) S^1 = COS $^{-1}$ [SINL/ Sin Lq + COS Lp COS Lq COS (Vp − Vq)
13) EL = TAN $^{-1}$ [(COS S^1 − R/r)/SIN S^1]
14) B = COS $^{-1}$ [(SIN Lp − SIN Lq COS S^1)/COS Lq SIN S^1]
15a) AZ = B (if SIN (Vp − Vq) ≤ 0)
15b AZ = 2 π − B (if SIN (Vp − Vq) · 0)

Note 1: If Lp 0 and I 90° or if Lp 0 and I · 90°
Note 2: If Lp · 0 and I · 90° or if Lp · 0 and I · 90°

**Not used in calculations, but of interest.

SPECIAL NOTES:

1) True anomaly, Ø should not be input higher than, or allowed to increment past 360 degrees in an attempt to follow a satellite through more than one complete orbit. Doing so will result in erroneous output.
2) New NBN times and NBN Longitudes should be input for each new orbit.

Table 6-7. WØSL Elliptical Orbit Program
(continued from page 80 and on page 82).

STEP	KEY ENTRY	KEY CODE	NOTES
001	*F LBL A	31 25 11	Start
	H CF 2	35 61 02	
	RCL 8	34 08	P
	RCL 9	34 09	Ø
	H π	35 73	
	G X ≤ Y	32 71	Is Ø ≧ 180 degrees?
	H SF2	35 51 02	
	H Rv	35 53	
	F COS	31 63	
010	RCL E	34 15	E
	X	71	
	1	01	
	+	61	
	÷	81	
	STO C	33 13	= r
	RCL 6	34 06	a
	÷	81	
	CHS	42	
	1	01	
020	+	61	
	RCL E	34 15	E
	÷	81	
	F GSB 6	31 22 06	(COS^{-1})
	F GSB 8	31 22 08	(2 π)
	H X ≤ Y	35 52	
	H F?2	35 71 02	Was Ø ≥ 180 degrees?
	−	51	= E
	Ent ↑	41	
	F SIN	31 62	
030	RCL E	34 15	E
	X	71	
	−	51	
	RCL 1	34 01	P
	F GSB 8	31 22 08	(2 π)
	÷	81	
	X	71	= T
	RCL 0	34 00	J
	−	51	
	STO 7	33 07	= T − J
040	6	06	
	0	00	
	÷	81	
	RCL A	34 11	NBN Time
	+	61	
	2	02	
	4	04	
	G X> Y	32 81	Is Time < 2400 Hrs?
	GTO 1	22 01	
	−	51	
050	GTO 2	22 02	
	* F LBL 1	31 25 01	
	H X⇄Y	35 52	

Table 6-7. WØSL Elliptical Orbit Program
(continued from page 81 and on page 83).

STEP	KEY ENTRY	KEY CODE	NOTES
	F X < 0	31 71	Is time < 0000 hrs?
	+	61	
	* F LBL 2	31 25 02	
	G→H. Ms	32 74	
	H ST I	35 33	= clock time of AZ-EL
	4	04	
	H 1/X	35 62	
060	G→R	32 73	
	STO X7	33 71 07	= (T − J) π/720
	RCL 2	34 02	I
	1	01	
	G SIN⁻¹	32 62	
	H CF3	35 61 03	
	G X > Y	32 81	Is I < 90 degrees?
	H SF3	35 51 03	
	H R↓	35 53	
	F SIN	31 62	
070	RCL 9	34 09	Ø
	F COS	31 63	
	CHS	42	
	X	71	
	G SIN⁻¹	32 62	
	STO B	33 12	= Lp (SSP latitude)
	FX < O	31 71	Is Latitude negative?
	H SF2	35 51 02	
	RCL D	34 14	Vx
	STO +7	33 61 07	= Vx + (T − J) π/720
080	RCL 9	34 09	Ø
	F SIN	31 62	
	RCL B	34 12	Lp
	F COS	31 63	
	÷	81	
	H F?2	35 71 02	was Latitude negative?
	H F?2	35 71 02	
	H F?3	35 71 03	was I · 90 degrees?
	H F?3	35 71 03	
	GTO 3	22 03	
090	CHS	42	
	H π	35 72	
	STO +7	33 61 07	= Vx + (T − J) π/720 +
	H X⇄Y	35 52	
	* F LBL 3	31 25 03	
	F GSB 6	31 22 06	(COS ⁻¹)
	STO + 7	33 61 07	= Vp (SSP longitude)
	RCL 7	34 07	
	F GSB 8	31 22 08	(2)
	G X ≤ Y	32 71	
100	STO −7	33 51 07	
	RCL 4	34 04	Vq
	STO −7	33 51 07	= (Vp − Vq)
	RCL 7	34 07	
	F COS	31 03	Lq

Table 6-7. WØSL Elliptical Orbit Program
(continued from page 82 and on page 84).

STEP	KEY ENTRY	KEY CODE	NOTES
	F COS	31 63	
	X	71	
	RCL B	34 12	Lp
	F COS	31 63	
110	X	71	
	RCL B	34 12	
	F SIN	31 62	
	RCL 3	34 03	Lq
	F SIN	31 62	
	X	71	
	+	61	
	F GSB 6	31 22 06	(COS^{-1})
	STO 5	33 05	= S^{-1}
	F COS	31 63	
120	6	06	
	3	03	
	7	07	
	1	01	
	R CL C	34 13	R
	÷	81	
	–	51	
	RCL 5	34 05	S^1
	F SIN	31 62	
	G→P	32 72	
130	H R↓	35 53	
	F D←	31 73	=EL
	F X<O	31 71	Is EL < 0?
	H SF2	35 51 02	
	DSP 2	23 02	
	F RND	31 24	
	ABS	35 64	Set EL = .EL
	1	01	
	F %	31 82	
	STO C	33 13	= .EL
140	RCL B	34 12	Lp
	F SIN	31 62	
	RCL 3	34 03	Lq
	F SIN	31 62	
	RCL 5	34 05	S^1
	F COS	31 63	
	X	71	
	–	51	
	RCL 3	34 03	Lq
	F COS	31 63	
150	RCL 5	34 05	S^1
	F SIN	31 62	
	X	71	
	F X ≠ O	31 61	(÷ by 0 protection)
	÷	81	
	F GSB 6	31 22 06	(COS^{-1})
	STO 5	33 05	= B
	CLX	00	

Table 6-7. WØSL Elliptical Orbit Program
(continued from page 83 and on page 85).

STEP	KEY ENTRY	KEY CODE	NOTES
	RCL 7	34 07	(Vp − Vq)
	F SIN	31 62	
160	G X≤Y	32 71	
	GTO 5	22 05	
	F GSB 8	31 22 08	(2π)
	RCL 5	34 05	B
	−	51	
	STO 5	34 05	$2\pi - B$
	* F LBL 5	31 25 05	
	H RCI	35 34	clocktime of AZ-EL
	DSP4	23 04	
	R/S	84	Display clocktime
170	RCL 5	34 05	B
	F D←	31 73	= AZ
	DSP O	23 00	
	F RND	31 24	
	DSP 2	23 02	
	RCL C	34 13	.EL
	+	61	
	H F?2	35 71 02	Was EL < 0?
	CHS	42	
	R/S	84	= AZ.EL display
180	1	01	
	0	00	
	G→R	32 73	
	STO +9	33 61 09	increment Ø by 10 degrees
	RCL 9	34 09	Ø
	F D←	31 73	
	H PAUSE	35 72	Display new Ø.
	GTO A	22 11	
	* F LBL 6	31 25 06	Subroutine for
	1	01	protecting against
190	G X≤Y	32 71	taking COS^{-1} of
	GTO 7	22 07	values of X where
	CHS	42	$-1 > X > 1$.
	GX=Y	32 71	
	HX⇆Y	35 52	
	* F LBL 7	31 35 07	
	G COS^{-1}	32 63	
	H TRN	35 22	
	* F LBL 8	31 25 08	subroutine for
	H π	35 73	obtaining altitude
200	ENT↑	41	
	+	61	
	H RTN	35 22	
	* F LBL C	31 25 13	Enter NBN time
	F H←	31 74	
	STO A	33 11	= NBN time
	G H.MS	32 74	
	DSP 4	23 04	
	H RTN	35 22	
	* F LBL D	31 25 14	Enter NBN longitude

Table 6-7. WØSL Elliptical Orbit Program
(continued from page 84).

Key		Code		Comment
G→R		32	73	
STO D		33	14	= NBN longitude
GTO 9		22	09	
* F LBL E	31	25	15	Enter true anomaly
G→R		32	73	
STO 9		33	09	= Ø
* F LBL 9	31	25	09	
F D←		31	73	
DSP 1		23	01	
H RTN		35	22	

K6PGX Successive Orbit Calculation Scheme

Dr Norm Chalfin, K6PGX, developed a straightforward yet effective method for determining orbital times for any satellite in circular orbit. The method is applicable to practically all pocket calculators, thus many amateurs should find it helpful. The following parameters will be assumed for Dr. Chalfin's "working example" which follows:

OSCAR 8 Satellite
Period: 103.23162 minutes
° = Degrees
Delta L° 25.80867° West

The reference orbit for March 29th (#325A) indicated EQX (equatorial crossing) at 0022.22 (00 hours 22 minutes and decimal minutes) UTC at longitude 44.40 degrees west. Those of you who have simple pocket digital calculators can readily compute successive EQXs from this data. Here's how:

1. Clear the calculator.
2. Enter longitude of EQZ – 44.40 for orbit 325A.
3. Press (+) and enter 25.81, press (=). The next EQX will be 70.21 for orbit 326A.
4. Press the (=) again. EQX is 96.02 for orbit 327A.
5. Press the (=) again. EQX is 121.83 for orbit 328A.

Each time the (=) is pressed thereafter the EQX will increment 25.81 degrees to give the next EQX. The 13th time you press (=) will indicate 379.93 degrees. At this point you subtract 360°. This gives you the EQX of the next orbit which is No. 388A at 19.93 degrees.

Now, press **CE**, or the clear entry key. Again you must press (+) and enter 25.81; repeat the successive operation of the (=) key until the next value in excess of 360° comes up, from which you will subtract the 360°. This will now be 14 more orbits, or orbit No. 352, and will be at 21.27°.

Again cancel the entry and repeat the process. If the day is Monday, Tuesday, Thursday or Friday, the orbit will be No. XXX-A; if Wednesday, No. XXX-X; if Saturday or Sunday, it will be No. XXX-J.

This gives you a list of EQXs at each orbit. Now you must increment the crossing times.

Since each orbit is 103.23 minutes long, you must add that to the starting time of 0022.22 for orbit No. 325 in order to obtain the EQX time for orbit No. 326. Note that the time increment for each orbit is one hour and 43.23 minutes. The crossing time for orbit No.

Table 6-8. Calculator Register Data for WØSL Elliptical Orbit Program.

0 = Time sine NBN (J)
1 = Period (P)
2 = Inclination (L)
3 = STA. latitude (Lq)
4 = STA. longitude (Vq)
4 = (used)
6 = a
7 = (used)
9 = P
9 = Ø

A = NBN time
B = (used)
C = (used)
D = (NBN longitude)
E = E
I = (used)

Calculator labels for previous program

A = Start
B = Not used
C = Enter NBN time
D = Enter NBN longitude
E = Enter true anomaly

Labels 1 through 9 used by program

Additional calculator notes

Trig. function set at rad.
Display function set at fixed :N2
Flags Ø through 3 set at off.

Table 6-9. Sample Problem for Use with Elliptical Orbit Program.

Hypothetical data of Phase III used in example.

R = 6371 Km
I = 101.7 degrees = 1.774999849 radians
hp = 1460 Km
P = 660 minutes
J = 32.8 minutes
Lq = 338.61260331 degrees = .673917061 radians
Vq = 90.51447368 degrees = 1.579775586 radians
a ≈ 25124.46863 Km
p = 13221.16983 Km
E = .688311816

Example: NBN at 105.5 degrees west at 06:31:05 GMT.
Begin at Ø = 70 degrees.

Storage required:

STO 0 = J = 32.8
STO 1 = P = 660
STO 2 = I = 1.774999849
STO 3 = Lq = .672917061
STO 4 = Vq = 1.579775586
STO 6 = a = 25124.46863
STO 8 = p = 13221.16983
STO E = E = .688311816

Solution:

Input		Output
105.5 [D]	→	105.5 (NBN longitude)
70 [E]	→	70.0 (Selected true anomaly)
06.3105 [C]	→	6.3105 (NBN clocktime)
[A]	→	6.1947 (AZ-EL clocktime)
[R/S]	→	– 189.05 (AZ = 189 degrees, EL = –5 degrees)
[R/S]	→	80.00** (incremented true anomaly)
	→	6.2450 (AZ-EL clocktime)
[R/S]		195.80 (AZ = 195 degrees, EL = 8 degrees) (etc.)

326 is 0205.45. The orbital crossing time at the equator for orbit No. 327 is 0348.68 UTC and so on.

Using the above information, you should be able to set up your own ephemerides (orbital equatorial crossing data).

Chapter 7 Satellite Equipment

Unlike many specialized areas of amateur radio, the OSCAR satellite enthusiast has a wide choice of commercial equipment at his disposal. This gear ranges from simple and inexpensive units to highly sophisticated transceivers with countless options. Mass production of amateur equipment has resulted in prices which are difficult to rival by homebrewing. Likewise, our ever rising economy creates the feasible possibility of purchasing a piece of new gear, using it for one or two years, then reselling it for almost the same amount of money paid for that unit. Amateur gear is, indeed, beginning to prove its worth as an inflation hedge.

One aspect which should not be overlooked by the amateur considering new gear is station compatibility. If, for example, the station's low band tranceiver uses a stout 12 volt power supply, this same supply can be used to power UHF tranceivers and linear amplifiers used for OSCAR communications. This power distribution also can be intermixed for mode A operations. For example, a 200 watt HF transceiver pulls miniscule current during receive, thus allowing its extra current to be used to power UHF amplifiers. Likewise, the gain of large antenna systems can be reduced by removing and storing some of their director elements. These elements can be reinstalled later, when satellite signals deteriorate.

The amateur satellite program is established in such a manner as to reuse specific bands with each new satellite. Thus, amateurs investing in satellite gear are situated in an advantageous position for many years.

The information presented in this chapter was secured from many of the manufacturers presently serving the OSCAR satellite enthusiast. It is intended as an up-to-date overview of available equipment, which you can relax and browse through while planning your own OSCAR setup. We hope you find the information interesting and helpful, and we look forward to hearing yoū via satellite in the near future.

YAESU

Yaesu Electronics Corp.
15954 Downey Ave.
Paramount, CA 90723

FT-225RD All-Mode Transceiver

Features

- **Full coverage.** The FT-225RD provides operation on all modes—SSB, CW, AM, and FM—over the entire 4 MHz of the 2 meter band.
- **PLL Circuitry.** The local oscillator employs the modern Phase Locked Loop (PLL) technique, with its fundamental oscillating in the 130 MHz range. This effectively eliminates spurious radiation and yields a clean output signal. On receive, the PLL likewise rejects unwanted out-of-band interference.
- **Frequency Memory Option.** Following the design philosophy pioneered in the FT-901DM HF transceiver, an available option for the FT-225RD is a memory unit which allows storage and recall of any frequency within the range of the transceiver. This circuitry allows instant, programmable QSY to a favorite repeater or calling frequency with only a flick of a switch.
- **Digital Plus Analog Frequency Readout.** The digital display uses large, bright LED's for maximum readability, with resolution to 0.1 kHz. The front panel lights and display may be dimmed, too, for nighttime mobile operation.
- **Versatility Features.** Squelch, VOX, PTT, semi-break-in CW with sidetone, and tone burst are standard features on the FT-225RD. A superb noise blanker permits mobile SSB operation, and a discriminator center meter allows precise zeroing on FM signals. The clarifier produces ±3.5 kHz offset of either the receive or trans-

Fig. 7-1. Yaesu FT-225RD 2 meter all-mode transceiver (courtesy Yaesu Electronics Corp.).

ceive frequency during VFO, memory, or fixed channel operation.

- **Fixed Channel Operation.** Up to 44 simplex or repeater channels may be installed through insertion of optional crystals in the FIX unit. These crystals are available through your Yaesu dealer.
- **AC/DC Capability.** The FT-225RD may be operated from 13.5 VDC, or from AC voltages of 100/110/117/200/220/234. Choice of AC or DC power is made by connection of the appropriate power cable to the receptacle on the rear panel of the transceiver.
- **Solid-State Modular Construction.** Yaesu's renowned plug-in circuit boards are utilized in the FT-225RD, providing maximum reliability and ease of servicing. All circuits are fully solid-state, using IC's and FET's for maximum performance capability within a compact case.

General Specifications

- **Frequency Range.** 144 ~ 145 MHz 146 ~ 147 MHz 145 ~ 146 MHz 147 ~ 148 MHz
- **Frequency Readout.** Digital readout to 0.1 kHz, analog display resolution better than 1 kHz.
- **Modes of Operation.** LSB, USB, CW, AM, FM.
- **Frequency Stability.** Within 100 Hz during any 30 minutes period after warmup. Not more than 20 Hz with 10% line voltage variation.

- **Intermediate Frequencies.** First IF = 10.7 MHz. Second IF = 455 kHz.
- **Antenna impedance.** 50 ohms unbalanced.
- **Repeater Split.** 600 kHz installed, any split up to 1 MHz possible with optional crystal.
- **Power Requirements.** AC 100/110/117/200/220/234 Volts, 50/60 Hz. DC 11.5—16.0 Volts, negative ground.
- **Power Consumption.** AC Receive 30 VA Transmit 160 VA at full output power DC Receive 1.2 A Transmit 6.5 A
- **Size.** 280(W) × 125(H) × 315(D) m/m.
- **Weight.** Approx. 9.0 kg.

Receiver Specifications

- **Sensitivity.** SSB/CW 0.5μV for 20 dB S/N FM 0.35 μV for 20 dB QS AM 1.0μV for 10 dB S/N (400 Hz 30% modulation)
- **Sensitivity.** SSB/CW/AM 2.3 kHz at 6 dB down 4.1 kHz at 60 dB down FM 12 kHz at 6 dB down 28 kHz at 60 dB down
- **Image Response.** Better than—60 dB
- **Spurious Response.** Better than 1μV at antenna input.
- **Audio Output Imepdance.** 4 ohms nominal.
- **Audio Output.** 2 watts at 10% THD

Transmitter Specifications

- **Audio Response.** 350 ~ 2600 Hz ±3 dB
- **Carrier Suppression.** Better than −40 dB
- **Spurious Radiation.** Better than −60 dB
- **FM Deviation.** Factory set at ±5 kHz, maximum 12 kHz
- **Power Output.** SSB 24 watts PEP FM/CW 25 watts AM 8 watts

FTV-901 VHF/UHF Transverter

The FTV-901 Transverter (Fig. 7-2) for the FT-901 series of transceivers provides transceive or split operation on the 50, 144, and 430 MHz bands, on all modes possible with the FT-901DM: SSB, CW, FSK, AM, and FM. Repeater offset of 1 MHz for 6 meters and 600 kHz for 2 meters is included. Power output is approximately 10 watts on all three bands.

Three satellite bands allow duplex operation on OSCAR modes A, B, and J, in conjunction with an external receiver. You can transmit on 145 and listen on 29 or 435 MHz, or transmit on 435

MHz and listen on 145 MHz. Receiving is full duplex, so you can monitor your downlink signal while transmitting.

Memory operation using the FT-901DM or FV-901DM memory system adds flexibility. On 6 meters, for example, a favorite beacon or calling frequency may be stored for quick reference. For satellite operation, the FV901DM scanner will quickly scan the downlink for incoming signals, and the frequency of schedules stations may be stored in memory.

The FTV-901 will be available soon, providing 9 bands of operation, from 160 meters to 70 cm, from your desk top.

Frequency coverage.

50 – 54 MHz	SAT-1: TX145 MHz, RX 29 MHz
144 – 148 MHz	SAT-2: TX 432 MHz, RX 145 MHz
430 – 440 MHz	SAT-3: TX 145 MHz, RX 435 MHz

SAT-1/2/3 bands require external receiver for duplex monitoring of downlink.

IF output. 28 – 30 MHz for FT-901 DM and external receiver.

Transmitter output. Approximately 10 watts.

Fig. 7-2. Yaesu FTV-901 VHF/UHF transverter.

Fig. 7-3. 432 MHz transverter (courtesy Amateur Radio Components Service).

ARCOS

Amateur Radio Components Service
P.O. Box 546
East Greenbush, NY 12061

Transmitting Converters

Originally intended for uplinking to the OSCAR satellites, these transmitting converters offer an economical and effective approach to SSB on 144, 220 and 432 MHz bands (Fig. 7-3).

Circuitry is simple and straight-forward. A solid state LO chain is followed by a 6939 mixer and a 6939 amplifier.

Proven, well tested design combined with rugged construction and the best available parts. Once peaked up after 10 minutes warm-up, will work indefinitely without further attention.

BNC Input and output connectors standard. Other types on request. LO mixing frequency available at $6.00 on initial order.

Output power. (nominal)
144MHz: 8 watts
220MHz: 6 watts
432MHz: 4 watts
Power required. 6V.AC at
3A.250V.DC at .14A.

Strip Line Power Amplifiers

The 420-450 MHz amplifier has parallel 4CX250's, 4CX350's or 8930's in AB1 linear operation. Quality parts. Chassis parts have welded seams with iridite finish. Stripline parts silver plated. Connecting cables and receptacles for power supply are furnished. Compact: 12″W×6″H×8″D. 600 watts output at 20 watts drive. Power supply requirements:

Eplate	2000V at ½A
Escreen	350V regulated
Egrid	60-70V reg./adj.
Efilament	5.6VAC at 6A

The 144-148 MHz amplifier has the same tube complement and description as above. Same power supply requirements (6VAC). Drive: 6 watts for 600 watts output. Attractive for use with the all-mode 2 M. transceivers on the market. Dual co-ax relays and control devices and information on reducing drive power and/or lowering amplifier gain to suit various models is available.

Both amplifiers have a similar appearance to the 70 CM amplifier in Fig. 7-4. All use common chassis parts. A 50 MHz amplifier built on the same chassis can also be furnished. Individual parts are available.

A wide choice of tubes is available. The 8930's have a larger anode surface—providing superior cooling characteristics. All of the 250 series perform well. The 150's are also ok-operated within their ratings. Orders must indicate if 8930's are the choice since the plate

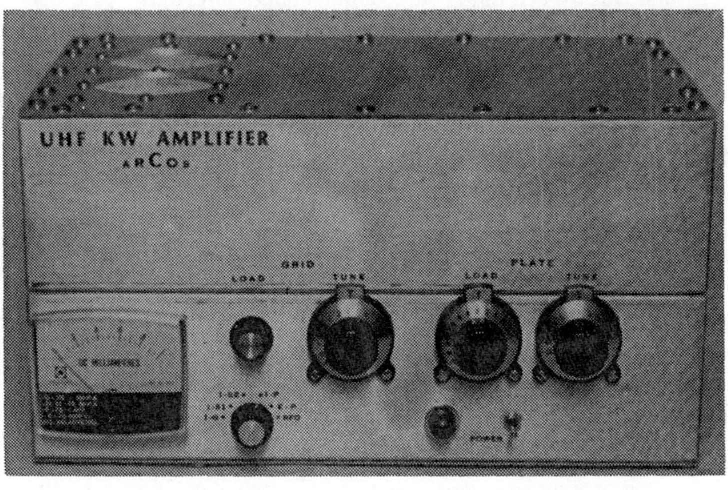

Fig. 7-4. 70 CM amplifier (courtesy Amateur Radio Components Service).

Fig. 7-5. Cushcraft 435 MHz 20-element twist antenna, right; 145 MHz 10-element beam, left (courtesy Cushcraft Coporation).

line, top cover and chimneys are dimensioned differently for these tubes.

If your needs include reliable VHF/UHF high power operation with a low drive power requirement, these amplifiers are your best choice. Make 'em yourself—the parts kits are supplied with all difficult mechanical and soldering operations completed—10 to 12 hours of pure assembly work.

CUSHCRAFT

Cushcraft Corporation
P.O. Box 4680
Manchester, NH 03108

Twist Antennas

Cushcraft offers complete antenna systems for OSCAR satellite communications on 2 meters and 70 cm. On 145 MHz the

10-element twist antenna (5 vertical elements, 5 horizontal) yields up to 10.8 dBd gain; the 20 element 435-MHz twist (10 vertical elements plus 10 horizontal) provides up to 13.6 dBd gain (Fig. 7-5). For greater performance on 2 meters choose the 20-element 145-MHz twist which offers 13.6 dBd gain. The half-power beamwidth of these antennas has been optimized for reliable satellite communications with minimum tracking requirements. All twist antennas come complete with coaxial matching harness for selectable horizontal, vertical, or circular polarization. Match 50-ohm feedlines.

OSCAR 7	**Input**	**Output**
Mode A	145.85-145.95	28.4-29.5
Mode B	432.125-432.175	145.975-145.925
OSCAR 8		
Mode A	145.85-145.95	28.4-29.5
Mode J	435.10-435.20	145.90-146.00

	145-MHz 10-element	**145-MHz 20-element**	**435-MHz 20-element**
Model	*A144-10T*	*A144-20T*	*A432-20T*
Center frequency	145.9 MHz	145.9 MHz	433.6 MHz
2:1 VSWR bandwith	>2 MHz	>1.5 MHz	>3 MHz
Horizontal gain	9.6 dBd	11.1 dBd	11.1 dBd
E plane beamwidth	52°	38°	38°
H plane beamwidth	63°	48°	48°
Vertical gain	9.5 dBd	11.1 dBd	11.1 dBd
E plane beamwidth	63°	48°	48°
H plane beamwidth	52°	38°	38°
Circular gain	10.8 dBd	12.3 dBd	12.3 dBd
E plane beamwidth	52°	38°	38°
H plane beamwidth	52°	38°	38°
Front-to-back ratio	20 dB	20 dB	20 dB
Boom length	70″ (1.8m)	103″ (3.3m)	57″ (1.4m)
Longest element	40″ (1.0m)	40″ (1m)	14″ (36cm)
Wind area ft^2 (m^2)	0.74 (.07)	1.42 ft^2 (0.13 m^2)	0.372 (.03)
Weight	3.5 lb (1.6 kg.)	6.6 lb (3.0kg.)	3.5 lb (1.6kg.)
Maximum mast OD	1½″ (3.8cm)	1½″ (3.8cm)	2″ (5.0cm)

Fig. 7-6. LPA 2-15 amplifier kit (courtesy Hamtronics Inc.).

HAMTRONICS

Hamtronics, Inc.
182 Belmont Rd
Rochester, NY 14612

OSCAR Amplifier

- Operate linear or class C, depending on wiring—Use on SSB, FM, CE, etc.
- Rugged, VSWR protected circuitry
- Latest, state-of-the-art emitter ballasted, high gain transistors
- Critical parts of assembly, like transistor mounting, done at the factory
- Smaller sizes can be mounted on rear panel of A25 Cabinet
- Choice of duty cycles to suit your application
- Easily tuned to adjacent commercial bands

LPA 2-15 kit (Fig. 7-6) features 1-2 W drive; 15 W p.e.p. output (linear); 20 W output (class C); 50% duty cycle; 13.6 VDC/2 - 3 Amp; Solid-state T/R switching; 2½ × 5″ heatsink; Available for 6M, 2M, 220MHz.

VHF Transmitting Converter

Use this inexpensive recycled 10 or 11M SSB exciter (Fig. 7-7) on VHF.

- Linear Converter for SSB, CW, FM
- A fraction of the cost of other units
- Ideal for OSCAR or terrestrial oper.
- Versatile module, separate from receiving converter for split bands
- Use with low power tap on exciter or regular output with attenuator
- Information included to make simple atten. pads for any power
- Easy to build and align
- Built-in test points; no special test equipment required

RF Input: Less than 1 mW p.e.p. (100) mV into 50 ohms)

RF Output: 2W p.e.p. continuous on ssb or keyed cw; 2W at 50% duty cycle on FM or carrier

Input/Output Impedance: 50 ohms

Power Required: Regulated 13.6 Vdc at 400 mA

Oscillator Freq: 38.667 MHz (XV2-4)

Freq Sensitivity: Less than 60 Hz over voltage range of ±1 VDC

Typical drift at 144 MHz: Less than 1 kHz/hr

Ambient Temp: Designed for constant room temperature operation

Relative 116 MHz output: −65dB

Harmonic Output: −60dB

Other Spurious: −60 dB

IM Distortion: −30 dB at recommended drive level (SSB)

Balanced J-fet mixer and shielded double tuned circuits for low spurs. Designed to fit A25 Cabinet

Fig. 7-7. XV2 VHF transmitting converter (courtesy Hamtronics Inc.).

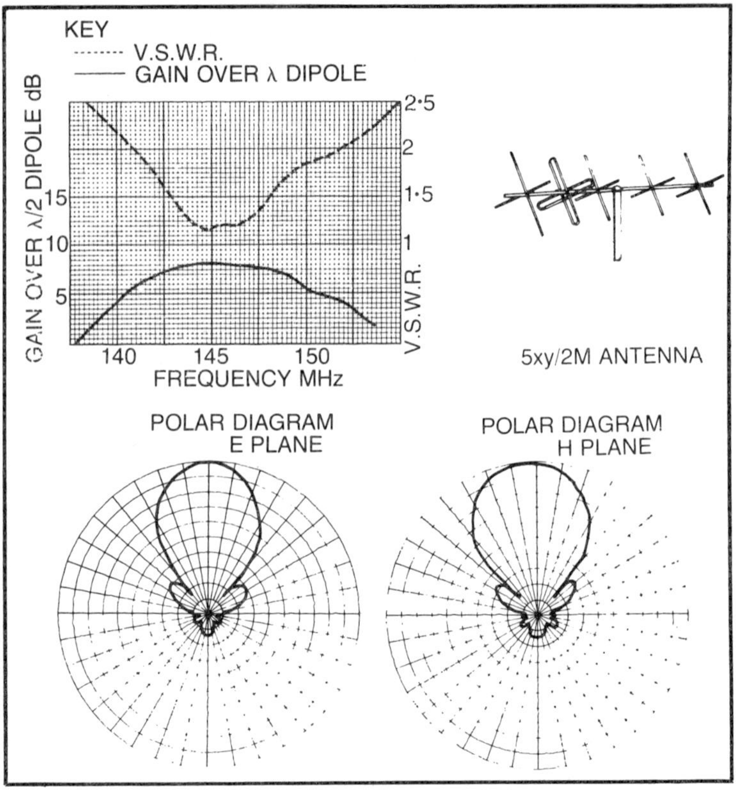

Fig. 7-8. Specifications for the 5XY/2 meter antenna (courtesy Spectrum International).

SPECTRUM INTERNATIONAL

Spectrum International, Inc.
P.O. Box 1084T
Concord, Mass. 01742

5XY/2 Meter Antenna

Gain: 7.8 dBd in each plane
Horizontal Beamwidth: 58°
Power Rating: 1 kW Peak
Weight: 2.8Kg
Wind Load at 160 Km/H: 19 Kg
Length: 1.7 meters
Design Impedance: Suitable for 50 ohms or 75 ohms

A separate harness Type PMH-2C is required for circular polarization. Also available for 134-138 MHz.

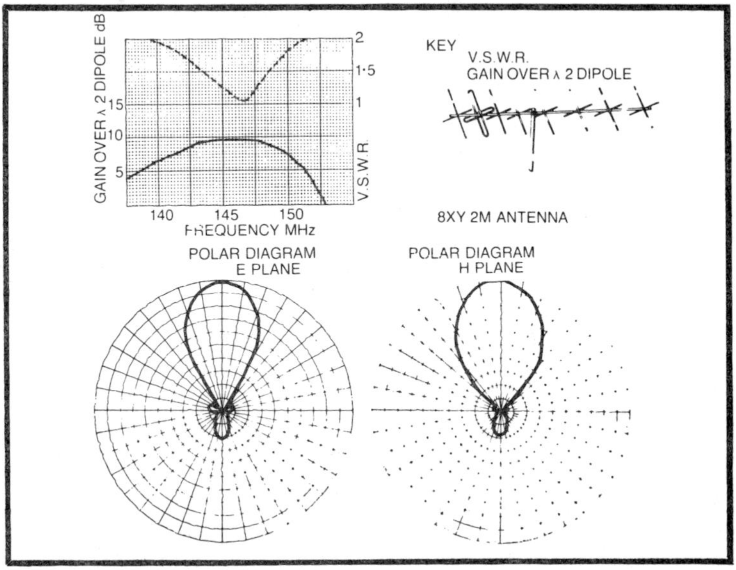

Fig. 7-9. Specifications for the 8XY/2 meter antenna (courtesy Spectrum International).

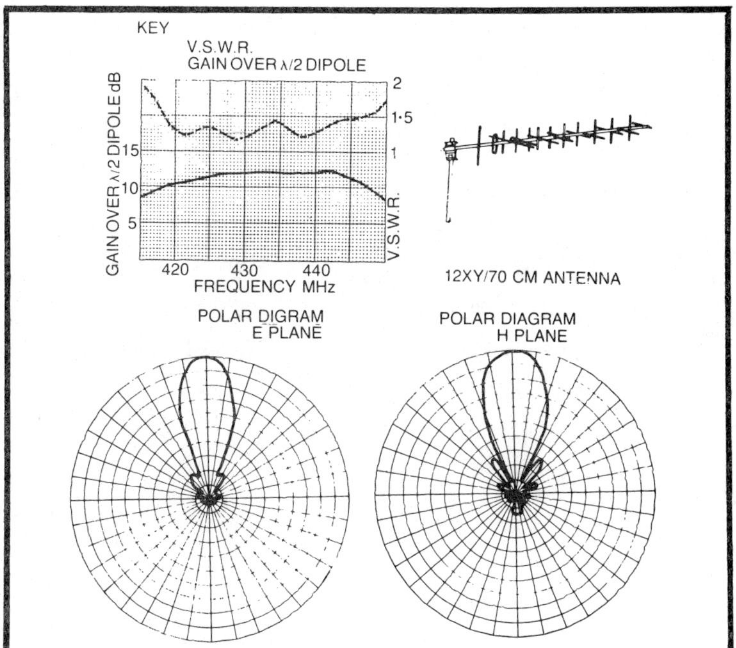

Fig. 7-10. Specifications for 12XY/70 cm antenna (courtesy Spectrum International).

8XY/2 Meter Antenna

Gain: 9.5 dBd in each plane
Horizontal Beamwidth: 47°
Power Rating: 1 Kw Peak
Weight: 4.7 Kg
Wind Load at 160 Km/h: 29 Kg
Length: 2.8 meters
Design Impedance: 50 ohms

A separate harness Type PMH-2C is required for circular polarization. Also available for 134-138 MHz.

12XY/70 cm Antenna

Gain: 13.0 dBd in each plane
Horizontal Beamwidth: 30°
Power Rating: (1 Kw Peak in each plane)
(110 watts peak via harness)

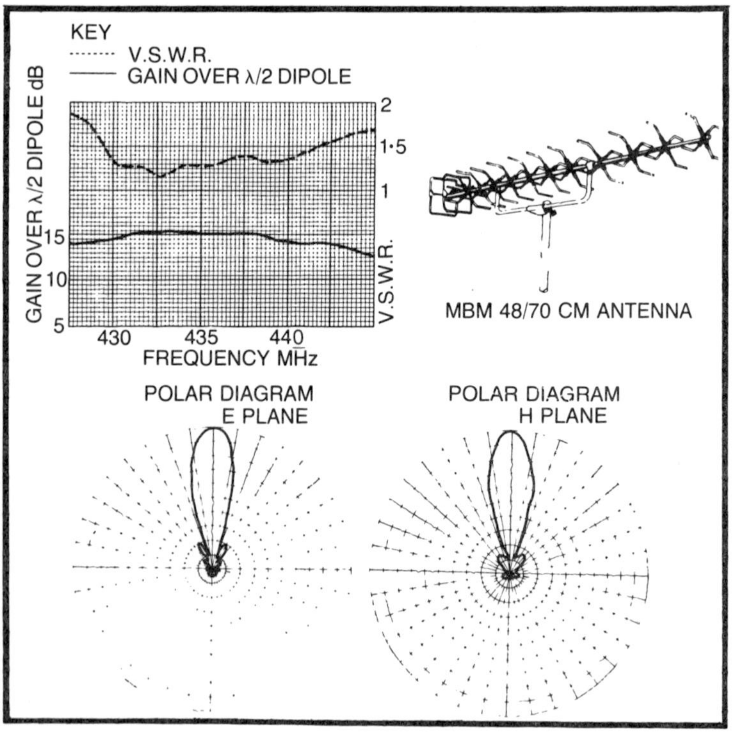

Fig. 7-11. Specifications for MBM 48/70 cm antenna (courtesy Spectrum International).

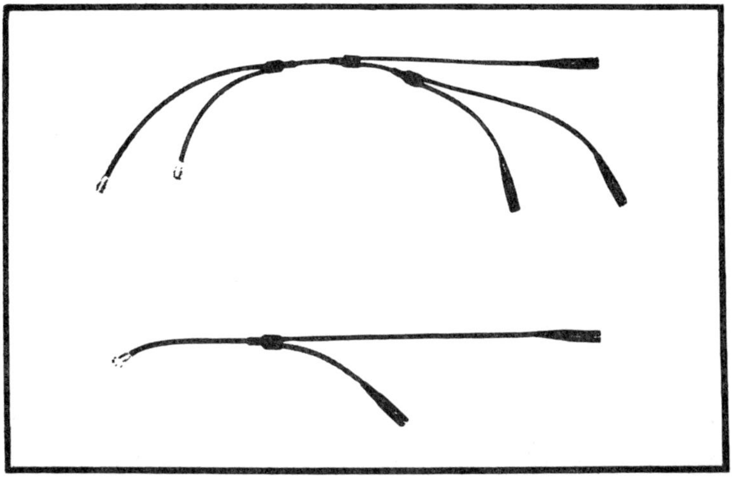

Fig. 7-12. Phasing harnesses. Two-way harness, left; four-way harness, right (courtesy Spectrum International).

Weight: 3.6 Kg
Wind Load at 160 Km/h: 21 Kg
Length: 2.6 meters
Design Impedance: 50 ohms

MBM48/70 cm Antenna

Gain: 15.7 dBd
Horizontal Beamwidth: 26°
Power Rating: 1 Kw Peak
Weight: 2.7 Kg
Wind Load at 160 Km/h: 17 Kg
Length: 1.83 meters
Design Impedance: 50 ohms

Phasing Harnesses

PMH2/4M 2 way phasing harness for two 4M yagis
PMH/2C 2 way phasing harness for circular polarization for two 2m aerials
SVMK/2M Mounting kit for vertical polarization for 2 slot-fed yagis
PMH2/2M 2 way phasing harness for two 2m aerials
PMH4/2M 4 way phasing harness for four 2m aerials
PMH2/70cm 2 way phasing harness for two 70cm aerials
PMH4/70cm 4 way phasing harness for four 70cm aerials

Glossary of OSCAR Terms

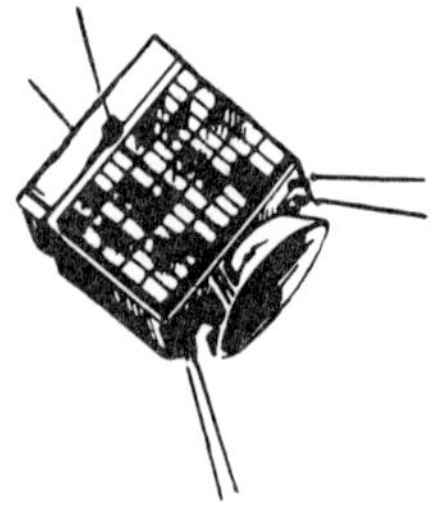

AMSAT—the Radio Amateur Satellite Corporation at P.O. Box 27, Washington, DC, 20044. This non-profit organization, in conjunction with the American Radio Relay League, coordinates all OSCAR satellite activities.

AOS—stands for Acquisition of Signal. The time that you can first hear signals from a specific satellite as it begins a pass near your location.

ascending node—a south or north pass of a specific satellite as viewed from Earth.

azimuth—beam headings of North, South, East, West, etc.

bandpass—frequency range of transponders used aboard OSCAR satellites.

descending node—a north to south pass of a specific satellite, as from earth.

doppler shift—phenomena which causes signals transmitted by a moving satellite to change frequency in accordance with the distance and speed of that satellite. All moving bodies exhibit this effect: train whistles are an excellent example.

downlink—the frequencies, or signals, transmitted from a satellite back to earth.

elevation—antenna tilt in degrees. Ninety degrees equals vertical position.

elliptical orbit—an eccentric, or off-center orbit which causes a satellite to move far into space during each daily pass or orbit.

ERP—stands for Effective Radiated Power. This level is the sum of transmitter output power and antenna gain. Example: 25 watt output into 6 db gain antenna = 100 watts ERP. Each 3 db doubles previous level.

frequency inversion—the effect of a satellite "swapping ends" on uplink and downlink bandpasses. This effect also causes USB signals to be retransmitted as LSB signals.

inclination—the angle of a particular satellite in reference to its equatorial crossing.

increment—number of degrees a satellite progresses west with each pass. This measurement is calibrated at the equator.

LOS—stands for Loss of Signal. The time you last hear signals from a specific satellite as it moves out of range of your location.

OSCAR—acronym for Orbital Satellite Carrying Amateur Radio. As of 1978, eight of these satellites have been placed in orbit around our earth.

pass—a single orbit of a particular satellite.

period—time of a complete orbit around the earth for a particular satellite.

polarization—angle of electromagnetic field of RF energy emanating from an antenna.

propagation delay—time required for signals to be relayed via satellite and returned to earth.

reference orbit—the first orbit, or pass, for a specific day after 0000 GMT.

QRP orbits—specific oribits, or passes, especially tailored to operations of 10 watts ERP or less.

telemetry beacon—transmitters aboard each OSCAR satellite which relay details on internal functions to earth in specially coded sequences.

transponder—name given to the linearly functioning repeater systems aboard a satellite.

tropospheric ducting—special form of UHF propagation which allows high frequency signals to be ducted through the troposphere, rather than continuing through it. Such ducted signals have been reported traveling several thousands of miles with minimal attenuation.

twist—special type of satellite antenna consisting of two yagi's mounted at right angles on a common boom. Phase relationships between these yagi's determines signal's direction of polarization.

uplink—the frequencies, or signals, transmitted from an earth-based station to a satellite.

Appendix A
AMSAT

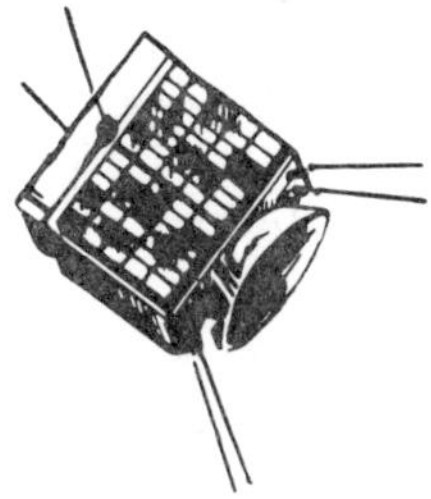

AMSAT, the Radio Amateur Satellite Corporation, P.O. Box 27, Washington, DC, 20044, was founded in 1969. The purposes and objectives of this corporation are as follows:

A. To provide satellites that can be used for amateur radio communication and experimentation by suitably equipped amateur radio stations throughout the world on a nondiscriminatory basis.

B. To encourage development of skills and the advancement of specialized knowledge in the art and practice of amateur radio communications and space science.

C. To foster international goodwill and cooperation through joint experimentation and study, and through the wide participation in these activities on a noncommercial basis by radio amateurs of the world.

D. To facilitate communications by means of amateur satellites in times of emergency.

E. To encourage the more effective and expanded use of the higher frequency amateur bands.

F. To disseminate scientific, technical and operational information derived from such communications and experimentation, and to encourage publication of such information in treatises, theses, trade publications, technical journals or other public media.

Membership in AMSAT is open to all radio amateurs and other interested persons. AMSAT encourages the participation of all interested individuals in its activities regardless of membership and invites licensed amateur radio operators of all countries to engage in radio transmissions to the satellite(s). Membership is possible in two categories:

AMSAT

Radio Amateur Satellite Corporation

P.O. BOX 27, WASHINGTON, D.C. 20044

(202)488-8649

AN INVITATION . . .

TO BECOME AN AMSAT LIFE MEMBER

For a contribution of $100 or more, you're invited to become an AMSAT Life Member. A bargain considering this is only 10 years' dues, and it's a tax-deductible donation. You may contribute in two or four installments if you prefer-- we only ask that you complete your payments by the end of the year. You can also charge your Life Membership to BankAmericard (VISA) or Master Charge.

As a Life Member,

- You receive the quarterly "AMSAT Newsletter" for life, now sent to Life Members by first class mail. If you're overseas, it's sent via airmail.

- We'll send you free, on request, AMSAT's W6PAJ orbital predictions book, which lists AMSAT-OSCAR satellite passes and operating schedules, annually.

- You'll receive other tokens of your support, such as an AMSAT-OSCAR pin, and a Life Membership certificate and card.

- Most importantly, your Life Membership helps sustain the amateur satellite program, making it possible for AMSAT to bring you newer and better satellites, serving as resources for the amateur community to use.

Won't you join us in Life Membership?

Fig. A-1. AMSAT life membership application (continued on page 109).

AMSAT

Radio Amateur Satellite Corporation

P.O. BOX 27, WASHINGTON, D.C. 20044

(202)488-8649

TO: AMSAT Life Membership Chairman

FROM: ______________________________
Name (print) Call

____
Street address

______________________ ____________
City, State (Country) ZIP/postal code

Yes, make me an AMSAT Life Member. I wish to contribute $________.

Method of payment:

Full payment enclosed ☐

Charge BankAmericard (VISA) ☐

Charge Master Charge ☐

______________________________ ____________
Charge card Account Number Expir. date

Payment in ☐two or ☐four installments.
Enclosed is $______ as first installment.
I agree to contribute $_______ on each of the following dates: ________________
____

____________ ________________
Date Signature

Fig. A-1. AMSAT life membership application (continued from page 108).

A. An interested individual may become a *Member* by filling out and returning the membership application, along with his dues payment (Fig. A-1).

B. A recognized group or organization interested in supporting AMSAT's goals and objectives and wishing to participate constructively in its activities may become a *Member Society* by completing and returning a member society application together with their dues payment. This class of membership was established to encourage interested groups to participate in AOSAT projects within the members society's own area of interest.

An annual financial contribution is requested from members and member societies in an effort to offset the costs of printing and mailing newsletters. This donation may be waived at the discretion of the Board of Directors. Donations are tax deductible.

Members of AMSAT are entitled to:

A. The opportunity to participate in the activities of AMSAT and to vote in the elections for the Board of Directors.
B. Receive newsletters and other information which may be generally distributed by AMSAT (Table A-1).
C. Be acknowledged as supporting the activities of AMSAT with a membership certificate or card.

Member Societies of AMSAT are entitled to:

A. Participate in AMSAT's acitivities.
B. Nominate two members per annum as candidates to the Board of Directors.
C. Receive newsletters and other information which may be distributed by AMSAT.
D. Be acknowledged as supporting AMSAT's activities with a Member Society card or certificate.

Since AMSAT's formation in 1969, this group has been responsible for several highly successful satellite projects. These projects include OSCAR 5, 6, 7, and 8. AMSAT is presently involved in the highly sophisticated Phase III satellite project (Fig. A-2) which will open new eras of UHF communications during late 1979 and the 1980's.

Table A-1. AMSAT Nets.

International Net	14280 kHz	Sundays 1800 GMT
East Coast U.S. Net	3850 kHz	Wednesdays 0100 GMT
Mid U.S. Net	3850 kHz	Wednesdays 0200 GMT
West Coast U.S. Net	3850 kHz	Wednesdays 0300 GMT

AMSAT

Radio Amateur Satellite Corporation

P.O. BOX 27, WASHINGTON, D.C. 20044

HELP TO REVOLUTIONIZE AMATEUR RADIO COMMUNICATIONS!

Here's a unique opportunity for you to become a part of an amateur satellite project. How? By sponsoring solar cells and other portions of the new AMSAT Phase III satellites.

An exciting new phase in amateur radio is about to begin, one that will affect all of us. OSCAR satellites of the new AMSAT Phase III series will soon revolutionize long-distance amateur communications in the same manner that earth-bound repeaters have completely transformed local communications -- by dramatically increasing communications reliability while simultaneously reducing the cost and complexity of individual amateur stations. The first Phase III spacecraft, now scheduled for launch in 1979, will be available to most stations about 17 hours each day, and will make communications possible between stations separated by distances of up to 11,000 miles.

Operators interested in DX, rag-chewing, contests and traffic handling will find Phase III satellites as easy to use as their favorite band. In effect, each satellite will provide a new band comparable to 20 meters, a resource usable by hundreds of stations at a time.

But your help is needed to make Phase III a success. Hardware costs for the Phase III project are expected to be $250,000. (A commercial satellite of similar capability would cost nearly $10,000,000). You can help by sponsoring one or more components in the new satellites with tax-deductible donations ranging from $10 to $10,000.

Won't you please help? Please return the enclosed sponsorship form with your contribution today. We'll send you a certificate acknowledging the specific components you are sponsoring.

Thank you for your support!

Fig. A-2. AMSAT Phase III sponsorship form (continued on page 112).

Radio Amateur Satellite Corporation

P.O. BOX 27, WASHINGTON, D.C. 20044

AMSAT-PHASE III SPONSORSHIP FORM

TO: AMSAT Phase III

FROM: Name__________________________ Call___________

Address____

____
City, State (Country)   ZIP

YES, I want to support AMSAT Phase III satellites. I wish to contribute $__________ in sponsorship of:

____ solar cells at $10 each
____ battery cells at $200 each
____ solar panels at $2,000 each
____ transponders at $5,000 each
____ rocket motor at $10,000

Method of payment:

☐ Check or money order enclosed

☐ Charge ☐ BankAmericard (VISA) or Master Charge ☐ Account Number: _________________________ Exp. date:______

☐ Payment in ☐ two or ☐ four installments. Enclosed is $______ as first installment. I agree to contribute $________ on each of the following dates:

____

Contributors of $1,000 or more will be honored by having their names inscribed on a plaque to be placed aboard the Phase III satellite for posterity. If you qualify, please indicate here how you would like your name and call listed:

____

Sponsorship donations are tax-deductible.

Date:______________ Signature:__________________

Fig. A-2. AMSAT Phase III sponsorship form (continued from page 111).

Appendix B Logs

SATELLITE LOG OF ____________

DATE	TIME	ORBIT	OSCAR#	STATION WORKED	HIS REPORT	MY REPORT	BAND MODE	ERP

Fig. B-1. Sample OSCAR operator's log.

SATELLITE LOG OF __________

DATE	TIME	ORBIT	OSCAR#	STATION WORKED	HIS REPORT	MY REPORT	BAND MODE	ERP	NOTES

Fig. B-1. Sample OSCAR operator's log.

SATELLITE LOG OF ____________

DATE	TIME	ORBIT	OSCAR#	STATION WORKED	HIS REPORT	MY REPORT	BAND MODE	ERP	NOTES

Fig. B-1. Sample OSCAR operator's log.

STATES WORKED VIA SATELLITE

STATE	CALL	DATE	TIME	ORBIT	OSCAR#	QSL	NOTES
Alabama							
Alaska							
Arizona							
Arkansas							
California							
Colorado							
Connecticut							
Delaware							
Florida							
Georgia							
Hawaii							
Idaho							

Fig. B-2. Sample operator's worked-all-states (WAS) list (continued on pages 117-119).

STATES WORKED VIA SATELLITE

STATE	CALL	DATE	TIME	ORBIT	OSCAR#	QSL	NOTES
Illinois							
Indiana							
Iowa							
Kansas							
Kentucky							
Louisiana							
Maine							
Maryland							
Massachusetts							
Michigan							
Minnesota							
Mississippi							

Fig. B-2. Sample states-worked-via-OSCAR log (continued from page 116 and on 118).

STATES WORKED VIA SATELLITE

STATE	CALL	DATE	TIME	ORBIT	OSCAR#	QSL	NOTES
Missouri							
Montana							
Nebraska							
Nevada							
New Hampshire							
New Jersey							
New Mexico							
New York							
North Dakota							
Ohio							
Oklahoma							
Oregon							

Fig. B-2. Sample WAS log (continued from page 117 and on 119).

STATES WORKED VIA SATELLITE

STATE	CALL	DATE	TIME	ORBIT	OSCAR#	QSL	NOTES
Pennsylvania							
Rhode Island							
South Carolina							
South Dakota							
Tennessee							
Texas							
Utah							
Vermont							
Virginia							
Washington							
West Virginia							
Wisconsin							
Wyoming							

Fig. B-2. Sample WAS log (continued from page 118).

SATELLITE EQUIPMENT LOG OF ______________________

(1) Date and location ______________________

Uplink equipment, antenna, ERP and SWR ______________________

____

____

____

Downlink equipment, antenna, filters, preamps, etc. ______________________

____

____

____

____

Changes made, their dates and results ______________________

____

____

____

____

(2) Second location and date ______________________

Uplink equipment, antenna, ERP and SWR ______________________

____

____

____

Downlink equipment, antenna, filters, preamps, etc. ______________________

____

____

____

Changes made, their dates and results ______________________

____

____

____

Fig. B-3. Sample satellite equipment log.

Appendix C

Phonetic Alphabet

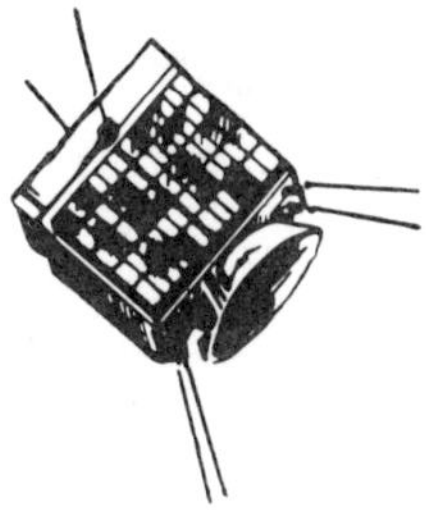

A ALFA
B BRAVO
C CHARLIE
D DELTA
E ECHO
F FOXTROT
G GOLF
H HOTEL
I INDIA
J JULIETTE
K KILO
L LIMA
M MIKE
N NOVEMBER
O OSCAR
P PAPA
Q QUEBEC
R ROMEO
S SIERRA
T TANGO
U UNIFORM
V VICTOR
W WHISKEY
X X-RAY
Y YANKEE
Z ZULU

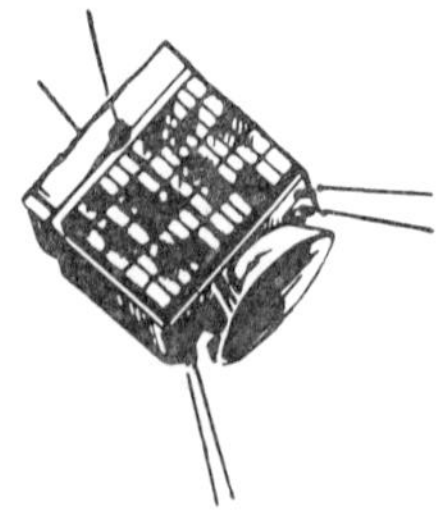

Appendix D International "Q" Signals

QRA	What is the name of your station?
QRB	How far are you from my station?
QRD	Where are you going and where are you from?
QRG	Will you tell me my exact frequency?
QRH	Does my frequency vary?
QRI	How is the tone of my transmission?
QRK	What is the readability of my signals?
QRL	Are you busy?
QRM	Are you being interferred with?
QRN	Are you troubled by static?
QRO	Shall I increase power?
QRP	Shall I decrease power?
QRQ	Shall I send faster?
QRR	Are you ready for automatic operation?
QRS	Shall I send more slowly?
QRT	Shall I stop sending?
QRU	Have you anything for me?
QRV	Are you ready?
QRW	Shall I inform ______ that you are calling him on__kHz.
QRX	When will you call me again?
QRY	What is my turn?
QRZ	Who is calling me?
QSA	What is the strength of my signals?
QSB	Are my signals fading?
QSD	Is my keying defective?

QSG Shall I send _____ telegrams at a time?

QSJ What is the charge to be collected per word to _____ including your internal telegraph charge?

QSK Can you hear me between your signals?

QSL Can you acknowledge receipt?

QSM Shall I repeat the last telegram which I sent you, or some previous telegram?

QSN Did you hear me on _____ kHz?

QSO Can you communicate with _____ direct or by relay?

QSP Will you relay to _____ free of charge?

QSQ Have you a doctor on board [or is ... (name of person) on board]?

QSU Shall I send or replay on this frequency [or on _____ kHz] (with emission of class_____)?

QSV Shall I send a series of V's on this frequency?

QSW Will you send on this frequency?

QSX Will you listen to _____ on _____ kHz?

QSY Shall I change to transmission on another frequency?

QSZ Shall I send each word or group more than once?

QTA Shall I cancel telegram number _____ as if it had not been sent?

QTB Do you agree with my counting of words?

QTC How many telegrams have you to send?

QTE What is my true bearing from you?

QTG Will you send two dashes of ten seconds each followed by your call sign [on _____ kHz]?

QTH What is your location?

QTI What is your true track?

QTJ What is your speed?

QTL What is your true heading?

QTN At what time did you leave _____ (place)?

QTO Have you left dock (or port)?
or
Are you airborne?

QTP Are you going to enter dock (or port)?
or
Are you going to land?

QTQ Can you communicate with my station by means of the International Code of Signals?

QTR What is the correct time?

QTS Will you send your call sign for _____ minute(s) on _____ kHz so that your frequency may be measured?

QTU What are the hours during which your station is open?

QTV Shall I stand guard for you on the frequency of _____ kHz?

QTX Will you keep your station open for further communication with me until further notice?

QUA Have you news of _____ (call sign)?

QUB Can you give me, in the following order, information concerning: visibility, height of clouds, direction and velocity (place of observation)?

QUC What is the number of the last message you received from me?

QUD Have you received the urgency signal, sent by _____ (call sign of mobile station)?

QUF Have you received the distress signal sent by ____ (call sign of mobile station)?

QUG Will you be forced to land?

QUH Will you give me the present barometric pressure at sea level?

Appendix E International Prefixes

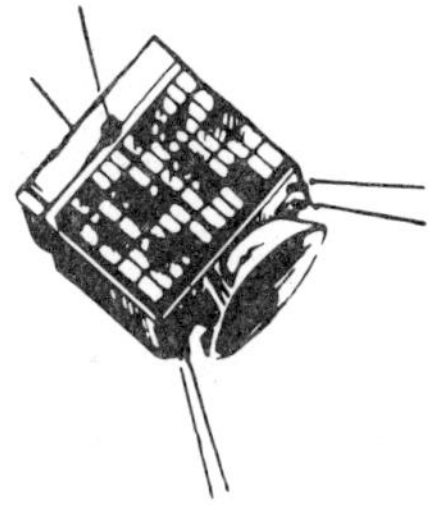

PREFIX	*COUNTRY*
AP	PAKISTAN
A2C	BOTSWANA
A35	TONGA
A4X	SULTANATE OF OMAN
A51	BHUTAN
A6X	UNITED ARAB EMIRATES
A7X	QATAR
A9X	BAHRAIN
BV	TAIWAN
BY	PEOPLES REPUBLIC OF CHINA
CE	CHILE
CE9AA-AM	CHILEAN ANTARCTICA TIERRA DE O'HIGGINS, PALMER PENINSULA, GRAHAM LAND
CE9AN-AZ	SOUTH SHETLAND IS.
CE0A	EASTER IS.
CE0X	SAN FELIX IS.
CE0Z	JUAN FERNANDEZ IS.
CM	CUBA
CN	MOROCCO
CO	CUBA
CP	BOLIVIA
CR3	GUINEA BISSAU
CR9	MACAO
CT1	PORTUGAL
CT2	AZORES IS.
CT3	MADEIRA IS.
CX	URUGUAY
C21	REPUBLIC OF NAURU
C31	ANDORRA
C5A	GAMBIA
C6A	BAHAMA IS.
C9M	MOZAMBIQUE
DA-DL	FED REP OF WEST GERMANY
DM	EAST GERMAN DEMOCRATIC REP.
DU	PHILIPPINES
D2A	ANGOLA
D4	REP OF CAPE VERDE
D6	STATE OF COMORO

EA .... SPAIN
EA6 .... BALEARIC IS.
EA8 .... CANARY IS.
EA9 .... SPANISH SAHARA, CEUTA Y MELILLA
EI .... IRELAND
EL,5L .... LIBERIA
EP .... IRAN
ET3 .... ETHIOPIA

F .... FRANCE
FB8W .... CROZET IS.
FB8X .... KERGUELEN IS.
FB8Y .... ANTARCTICA
FB8Z .... AMSTERDAM & ST. PAUL IS.
FC .... CORSICA
FG7 .... GUADELOUPE
FH8 .... MAYOTTE
FK8 .... NEW CALEDONIA
FM7 .... MARTINIQUE
FO8 .... FRENCH POLYNESIA, CLIPPERTON IS.
FP8 .... ST. PIERRE & MIQUELON IS.
FR7 .... GLORIOSO, JUAN DE NOVA, REUNION, TROMELIN IS.
FS7 .... ST. MARTIN IS.
FW8 .... WALLIS & FUTUNA IS.
FY7 .... FRENCH GUIANA

G .... ENGLAND
GD .... ISLE OF MAN
GI .... NORTHERN IRELAND
GJ .... JERSEY
GM .... SCOTLAND
GU .... BAILIWICK OF GUERNSEY
GW .... WALES

HA,HG .... HUNGARY
HB .... SWITZERLAND
HB0 .... LIECHTENSTEIN
HC .... ECUADOR
HC8 .... GALAPAGOS IS.
HG .... HUNGARY
HH .... HAITI
HI .... DOMINICAN REPUBLIC
HK .... COLOMBIA
HK0 .... BAJO NUEVO,MALPELO,SAN ANDRES & PROVIDENCIA IS.
HK0 .... SERRANA BANK & RONCADOR CAY
HM,HL9 .... KOREA
HP .... PANAMA
HR .... HONDURAS
HR0 .... SWAN IS.
HS .... THAILAND
HV .... VATICAN CITY
HZ,7Z .... SAUDI ARABIA
H4 .... SOLOMON IS.

I, IW .... ITALY
IA .... TUSCAN ARCHIPELAGO

IC .... CAPRI & ISCHIA IS.
IG9 .... LAMPEDUSA IS.
IH .... PANTELLERIA IS.
IM .... MADDALENA IS.
IS .... SARDINIA
IT .... SICILY

JA,JE-JJ,JR .... JAPAN
JD1 .... OGASAWARA,MINAMI-TORI-SHIMA IS.
JR6 .... OKINAWA (RYUKYU IS.)
JT1 .... MONGOLIA
JW .... SVALBARD IS.
JX .... JAN MAYEN IS.
JY .... JORDAN
J2 .... REP OF DJIBOUTI
J3 .... GRENADA & DEPENDENCIES

K .... UNITED STATES OF AMERICA
KA .... U.S.PERSONNEL IN JAPAN
KB6 .... BAKER,CANTON,ENDERBURY, HOWLAND & PHOENIX IS.
KC4 .... NAVASSA IS.
KC4AA,KC4US .... ANTARCTICA
KC6 .... CAROLINE IS.
KG4 .... GUANTANAMO BAY
KG6 .... MARIANA IS.
KG6 .... GUAM
KG6R .... ROTA
KG6S .... SAIPAN
KH6 .... HAWAII & KURE IS.
KJ6 .... JOHNSTON IS.
KL7 .... ALASKA
KM6 .... MIDWAY IS.
KP4 .... PUERTO RICO
KP6 .... JARVIS & PALMYRA IS.
KS6 .... AMERICAN SAMOA
KV4 .... VIRGIN IS.
KW6 .... WAKE IS.
KX6 .... MARSHALL IS.
K25 .... CANAL ZONE

LA-LJ .... NORWAY
LU .... ARGENTINA
LU-Z .... ANTARCTICA
LX .... LUXEMBOURG
LZ .... BULGARIA

M1 .... SAN MARINO

N .... UNITED STATES OF AMERICA

OA .... PERU
OD5 .... LEBANON
OE .... AUSTRIA
OH .... FINLAND
OHO .... ALAND IS.
OJO, OHOM .... MARKET REEF

OK,OL .... CZECHOSLOVAKIA
ON .... BELGIUM
OR .... ANTARCTICA
OX .... GREENLAND
OY .... FAEROES IS.
OZ .... DENMARK

PA-PI .... NETHERLANDS
PJ .... NETHERLANDS ANTILLES
PJ2,9 .... CURACAO
PJ3,9 .... ARUBA
PJ4,9 .... BONAIRE
PJ5,8 .... ST. EUSTATIUS
PJ6,8 .... SABA IS.
PJ7,8 .... SINT MAARTEN
PP-PY .... BRAZIL
PY0 .... FERNANDO DE NORONHA IS.
PY0 .... ST. PETER & ST. PAUL'S ROCKS
PY0 .... TRINIDADE & MARTIM VAZ IS.
PZ .... SURINAM
P29 .... PAPUA NEW GUINEA

SJ-SM .... SWEDEN
SP .... POLAND
ST .... SUDAN
SU .... EGYPT
SV .... CRETE,GREECE
SV5 .... DODECANESE IS.
S2,S3 .... BANGLADESH
S7 .... SEYCHELLES IS.
S8 .... TRANSKEI
S9 .... SAO TOME & PRINCIPE IS.

TA,TC .... TURKEY
TF .... ICELAND
TG .... GUATEMALA
TI .... COSTA RICA
TI9 .... COCOS IS.
TJ .... CAMEROON
TL8 .... CENTRAL AFRICAN REPUBLIC
TN8 .... REPUBLIC OF CONGO
TR8 .... GABON REPUBLIC
TT8 .... REPUBLIC OF CHAD
TU2 .... IVORY COAST
TY .... PEOPLLES REPUBLIC OF BENIN
TZ .... MALI REPUBLIC

UA1,2,3,4,6 .... EUROPEAN RUSSIAN SOVIET FEDERATED SOCIALIST REPUBLIC
UA9,0 .... ASIATIC RUSSIAN S.F.S.R.
UA1 .... FRANZ JOSEF LAND
UA2 .... KALININGRADSK
UB5 .... UKRAINIAN S.S.R.
UC2 .... WHITE RUSSIAN S.S.R.
UD6 .... AZERBAIDZHAN S.S.R.
UF6 .... GEORGIAN S.S.R.
UG6 .... ARMENIAN S.S.R.

UH8 ... TURKMEN S.S.R.
UI8 ... UZBEK S.S.R.
UJ8 ... TADZHIK S.S.R.
UL7 ... KAZAKH S.S.R.
UM8 ... KIRGHIZ S.S.R.
UN1 ... KARELO-FINNISH S.S.R.
UO5 ... MOLDAVIAN S.S.R.
UP2 ... LITHUANIAN S.S.R.
UQ2 ... LATVIAN S.S.R.
UR2 ... ESTONIAN S.S.R.

VE ... CANADA
VK ... AUSTRALIA
  VK2 ... LORD HOWE IS.
  VK9N ... NORFOLK IS.
  VK9X ... CHRISTMAS IS.
  VK9V ... COCOS IS.
  VK9Z ... WILLIS IS.
  VK0 ... ANTARCTICA
VO1 ... NEWFOUNDLAND
VO2 ... LABRADOR
VP1 ... BELIZE
VP2 ... LEEWARD & WINDWARD IS.
  VPZA ... ANTIGUA,BARBUDA
  VP20 ... DOMINICA
  VP2E ... ANGUILLA
  VP2K ... ST KITTS,NEVIS
  VP2L ... ST LUCIA
  VP2M ... MONSTERRAT
  VP25 ... ST VINCENT & DEPENDENCIES
  VP2V ... BRITISH VIRGIN IS.
VP5 ... TURKS & CAICOS IS.
VP8 ... FALKLAND,S.GEORGIA,S.ORKNEY, S.SANDWICH,S.SHETLAND IS.,GRAHAM LAND
VP9 ... BERMUDA IS.
VQ9 ... CHAGOS
VR1 ... BRITISH PHOENIX,GILBERT & OCEAN IS.
VR3 ... NORTHERN LINE IS.
VR6 ... PITCAIRN IS.
VR7 ... CENTRAL & SOUTHERN LINE IS.
VR8 ... TUVALU IS.
VS5 ... BRUNEI
VS6 ... HONG KONG
VU2 ... INDIA
  VU7 ... ANDAMAN & NICOBAR IS.
  VU7 ... LACCADIVE IS.

W ... UNITED STATES OF AMERICA

XE,XF ... MEXICO
  XF4 ... REVILLA GIGEDO IS.
XT2 ... VOLTAIC REPUBLIC
XU ... CAMBODIA/KHMER REPUBLIC
XV5 ... VIETNAM
XW8 ... LADS
XZ ... BURMA

YA .... AFGHANISTAN
YB-YD .... INDONESIA,TIMOR IS.
YI .... IRAQ
JQ .... NEW HEBRIDES
YK .... SYRIA
YN .... NICARAGUA
YO .... ROMANIA
YS .... EL SALVADOR
YU .... YUGOSLAVIA
YV .... VENEZUELA
    YVO .... AVES IS.

ZA .... ALBANIA
ZB2 .... GIBRALTAR
ZD7 .... SAINT HELENA IS.
ZD8 .... ASCENSION IS.
ZD9 .... TRISTAN DA CUNHA & GOUGH IS.
ZE .... RHODESIA
ZF1 .... CAYMAN IS.
ZK1 .... COOK & MANIHIKI IS.
ZK2 .... NIUE IS.
ZL .... NEW ZEALAND & AUCKLAND,CAMPBELL,CHATHAM,KERMADEC IS.
    ZL5 .... ANTARCTICA
ZM7 .... TOKELAU IS.
ZP .... PARAGUAY
ZR,ZS1,2,4,5,6 .... REPUBLIC OF SOUTH AFRICA
    ZS1ANT .... ANTARCTICA
    ZS2 .... PRINCE EDWARD & MARION IS.
ZR,ZS3 .... SOUTHWEST AFRICA (NAMIBIA)

3A .... MONACO
3B6 .... AGALEGA IS.
3B7 .... ST. BRANDON IS.
3B8 .... MAURITIUS IS.
3B9 .... RODRIGUEZ IS.

3C .... EQUATORIAL GUINEA
3D2 .... FIJI IS.
3D6 .... SWAZILAND
3V8 .... TUNISIA
3X .... REPUBLIC OF GUINEA
3Y .... BOUVET IS.

4K1 .... ANTARCTICA
4S7 .... SRI LANKA
4U .... UNITED NATIONS,GENEVA
4W .... YEMEN
4X4,4Z4 .... ISRAEL

5A .... LIBYAN ARAB REPUBLIC
5B4,ZC4 .... CYPRUS
5HI .... ZANZIBAR,TANZANIA
5H3 .... TANZANIA
5L .... LIBERIA
5N2 .... NIGERIA
5R8 .... MALAGASY REPUBLIC
5T5 .... MAURITANIA

5U7	NIGER
5V	TOGO
5W1	WESTERN SAMOA
5X5	UGANDA
5Z4	KENYA
6O	SOMALI REPUBLIC
6W8	SENEGAL REPUBLIC
6Y5	JAMAICA
7J	OKINO-TORI-SHIMA
7O	SOUTH YEMEN & KAMARAN IS
7P8	LESOTHO
7Q7	MALAWI
7X	ALGERIA
7Z	SAUDI ARABIA
8J	ANTARCTICA
8P6	BARBADOS IS.
8Q6	MALDIVE IS.
8R	GUYANA
8Z4	SAUDI ARABIA/IRAQ NEUTRAL ZONE
9G1	GHANA
9H1,5	MALTA
9H4	GOZO (MALTA)
9I,9J	ZAMBIA
9K2	KUWAIT
9L1	SIERRA LEONE
9M2	WEST MALAYSIA
9M6	SABAH
9M8	SARAWAK
9N1	NEPAL
9Q5	REPUBLIC OF ZAIRE
9U5	BURUNDI
9V1	REPUBLIC OF SINGAPORE
9X	RWANDA
9Y4	TRINIDAD & TOBAGO IS.

COUNTRY	*PREFIX*
AFGHANISTAN	**YA**
AGALEGA IS	**3B6**
ALAND IS	**OHO**
ALASKA	**KL7**
ALBANIA	**ZA**
ALGERIA	**7X**
AMSTERDAM & ST. PAUL IS	**FB8Z**
ANDAMAN & NICOBAR IS	**VU7**
ANDORRA	**C31**
ANGOLA	**D2A**
ANTARCTICA	**CE9AA-AM,FBBY,KC4,LU-Z, UA1,VKO,VP8,ZL5,ZS1ANT,3Y,4K1,8J**
ARGENTINA	**LU**
ARUBA	**PJ3,9**
ASCENSION IS	**ZD8**
AUCKLAND & CAMPBELL IS	**ZL**
AUSTRALIA	**VK**
AUSTRIA	**OE**

AVES IS.......YVO
AZDRES IS.......CT2

BAHAMA IS.......C6A
BAHRAIN IS.......A9X
BAJO NUEVO IS.......HKO
BAKER IS.......KB6
BALEARIC IS.......EA6
BANGLADESH.......S2,S3
BARBADOS IS.......8P6
BELGIUM.......ON
BELIZE.......VPI
BENIN, PEOPLE'S REP OF.......TY
BERMUDA IS.......VP9
BHUTAN.......A51
BOLIVIA.......CP
BONAIRE.......PJ4,9
BOTSWANA.......A2C
BOUVET IS.......3Y
BRAZIL.......PP-PY
BRITISH PHOENIX IS.......VRI
BRUNEI.......VS5
BULGARIA.......LZ
BURMA.......XZ
BURUNDI.......9U5

CAMBODIA/KHMER REP.......XU
CAMEROON.......TJ
CANADA.......VE
CANAL ZONE.......KZ5
CANARY IS.......EA8
CANTON IS.......KB6
CAPE VERDE, REP OF.......D4
CAPRI IS.......IC
CAROLINE IS.......KC6
CAYMAN IS.......ZF1
CENTRAL AFRICAN REPUBLIC.......TL8
CEUTA Y MELILLA, SPANISH.......EA9
CHAD REPUBLIC.......TT8
CHAGOS IS.......VQ9
CHATHAM IS.......ZL
CHILE.......CE
CHINA, PEOPLES REP OF.......BY
CHRISTMAS IS.......VK9X
CLIPPERTON IS.......FO8
COCOS IS.......TI9
COCOS (KEELING) IS.......VK9Y
COLUMBIA.......HK
COMORO,STATE OF.......D6
CONGO, REPUBLIC OF.......TN8
COOK IS.......ZK1
CORSICA.......FC
COSTA RICA.......TI
CRETE, GREECE.......SV
CROZET IS.......FB8W
CUBA.......CM,CO
CURACAO.......PJ2,9

CYPRUS 5B4,ZC4
CZECHOSLOVAKIA .... OK,OL

DENMARK .... OZ
DJIBOUTI,REP OF J2
DODECANESE IS SV5
DOMINICAN REPUBLIC .... HI

EASTER IS .... CEOA
ECUADOR .... HC
EGYPT .... SU
EL SALVADOR .... YS
ENDERBURY IS KB6
ENGLAND .... G
ETHIOPIA ET3

FAEROES IS .... DY
FALKLAND IS VP8
FERNANDO DE NORONHA .... PYO
FIJI IS 3D2
FINLAND .... OH
FRANCE .... F
FRENCH POLYNESIA FD8

GABON REPUBLIC TR8
GALAPAGOS IS HC8
GAMBIA C5A
GERMANY, FED. REP. (WEST) .... DA-DL
GERMAN DEM. REP. EAST .... DM
GHANA 9G1
GIBRALTAR ZB2
GILBERT & OCEAN IS .... VRI
GLORIOSO IS FR7
GOUGH IS ZD9
GOZO (MALTA) 9H4
GRAHAM LAND VP8,LU-Z
GREECE .... SV
GREENLAND .... OX
GRENADA J3
GUADELOUPE FG7
GUAM IS KG6
GUANTANAMO BAY KG4
GUATEMALA .... TG
GUERNSEY,BAILIWICK OF .... GU
GUIANA, FRENCH FY7
GUINEA BISSAU CR3
GUINEA, EQUATORIAL 3C
GUINEA, REPUBLIC OF 3X
GUYANA 8R

HAITI .... HH
HAWAII KH6
HEARD IS .... VKO
HONDURAS .... HR
HONG KONG VS6
HOWLAND IS KB6
HUNGARY .... HA, HG

ICELAND .... TF
INDIA VU2
INDONESIA .... YB-YD
IRAN .... EP
IRAQ Y1
IRELAND E1
ISCHIA IC
ISLE OF MAN .... GD
ISRAEL 4X4,4Z4
ITALY .... I,IW
IVORY COAST TU2

JAMAICA 6Y5
JAN MAYEN IS .... JX
JAPAN .... JA-JR
JAPAN, U.S.PERSONNEL IN .... KA
JARVIS IS KP6
JERSEY .... GJ
JOHNSTON IS KJ6
JORDON .... JY
JUAN DE NOVA IS FR7
JUAN FERNANDEZ IS CE02

KAMARAN IS 7D
KENYA 5Z4
KERGUELEN IS FB8X
DERMADEC IS .... ZL
KOREA HM,HL9
KURE IS KH6
KUWAIT 9K2

LABRADOR VD2
LACCADIVE IS VU7
LAMPEDUSA IS IG9
LAOS XW8
LEBANON DD5
LEEWARD IS.
    ANGUILLA .... VPZE
    ANTIGUA, BARBUDA VP2A
    BRITISH VIRGIN IS VP2V
    MONTSERRAT VP2M
    ST. KITTS, NEVIS VP2K
LESOTHO 7P8
LIBERIA EL,5L
LIBYAN ARAB REPUBLIC 5A
LIECHTENSTEIN .... HBO
LINE IS, NORTHERN VR3
LINE IS,CENTRAL & SOUTHERN VR7
LORD HOWE IS VK2
LUXEMBOURG LX

MACAO CR9
MACQUARIE IS .... VKO
MADDALENA IS .... IM
MADEIRA IS CT3
MALAGASY REPUBLIC 5R8
MALAWI 7Q7

MALAYSIA, EAST	9M6,8
MALAYSIA, WEST	9M2
MALDIVE IS	8Q6
MALI REPUBLIC	TZ
MALPELO IS	HK0
MALTA	9HI,5
MANIHIKI IS	2K1
MARIANA IS	KG6
MARKET REEF	OJ0,OH0M
MARSHALL IS	KX6
MARTINIQUE	FM7
MAURITANIA	5T5
MAURITIUS IS	36B
MAYOTTE	FH8
MEXICO	XE,AF
MIDWAY IS	KM6
MINAMI-TORI-SHIMA IS	JD1
MONACO	3A
MONGOLIA	JT1
MOROCCO	ON
MOZAMBIQUE	O9M
NAURU, REPUBLIC OF	Q21
NAVASSA IS	KC4
NEPAL	9N1
NETHERLANDS	PA-PI
NETHERLANDS ANTILLES	PJ
NEW CALEDONIA	FK8
NEWFOUNDLAND	VO1
NEW HEBRIDES	YJ
NEW ZEALAND	ZL
NICARAGUA	YN
NIGER	5U7
NIGERIA	5N2
NIUE IS	ZK2
NORFOLK IS	VK9N
NORTHERN IRELAND	GI
NORWAY	LA-LJ
OGASAWARA IS	JDI
OKINAWA (RYUKYU IS.)	JR6
OKINO-TORI-SHIMA	7J1
OMAN, SULTANATE OF	A4K
PAKISTAN	AP
PALMYRA IS	KP6
PANAMA	HP
PANTELLERIA IS	IH
PAPUA NEW GUINEA	P29
PARAGUAY	ZP
PERU	OA
PHILLIPINES	DU
PHEONIX IS	K36
PITCAIRN IS	VR6
POLAND	SP
PORTUGAL	CT1
PRINCE EDWARD & MARION IS.	Z52
PRINCIPE & SAO TOME IS	S9
PUERTO RICO	KP4

QATAR A7K

REUNION IS FR7
REVILLA GIGEDO IS XF4
RHODESIA .... ZE
RODRIGUEZ IS 3B9
ROMANIA .... YO
ROTA KG6R
RWANDA 9X

SABAH 9M6
SABA IS PJ6,8
ST. BRANDON IS 3B7
ST. EUSTATIUS PJ5,8
SAINT HELENA IS ZD7
SAINT MARTIN IS FS7
ST. PETER & ST. PAULS ROCKS .... PYO
ST. PIERRE & MIQUELON IS FP8
SAIPAN KG6S
SAMOA, AMERICAN KS6
SAMOA, WESTERN 5WI
SAN ANDRES & PROVIDENCIA .... HKO
SAN FELIX IS .... CEOX
SAN MARINO M1
SARAWAK 9M8
SARDINIA .... IS
SAUDI ARABIA .... HZ, 7Z
SAUDI ARABIA/IRAQ
NEUTRAL ZONE BZ4
SCOTLAND .... GM
SENEGAL REPUBLIC 6W8
SERRANA BK & RONCADOR CAY 8KO
SEYCHELLES IS S79
SICILY .... IT
SIERRA LEONE 9LI
SINGAPORE, REP OF 9V1
SINT MAARTEN PJ7,8
SOLOMON IS H4
SOMALI REPUBLIC 60
SOUTH AFRICA,REP OF ZR,AZQ,2,4-6
SOUTH GEORGIA IS VP8
SOUTH ORKNEY IS .... QZ, 2, VP8
SOUTH SANDWICH IS .... QZ, 2, VP8
S.SHETLAND IS CE9AN-AZ,QZ-Z,VP8
SOUTHWEST AFRICA(NAMIBIA) ZR3,ZS3
SOUTH YEMAN 70
SOVIET UNION:
EUROPEAN RUSSIA SOVIET FEDERATED
SOCIALIST REPUBLIC UA1,2,3,4,6
ASIATIC RUSSIAN SFSR UA9,0

ARMENIAN S.S.R UG6
AZERBAIDZHAN S.S.R UD6
ESTONIAN S.S.R UR2
FRANZ JOSEF LAND UA1
GEORGIAN S.S.R UF6
KALININGRADSK UA2
KARELO-FINNISH S.S.R UN1

KAZAKH S.S.R **UL7**
KIRGHIZ S.S.R .... **UMB**
LATVIAN S.S.R **UQ2**
LITHUANIAN S.S.R **UP2**
MOLDAVIAN S.S.R **UO5**
TADZHIK S.S.R **UJ8**
TURKMEN S.S.R **UH8**
UKRANIAN S.S.R **UB5**
UZBEK S.S.R **UI8**
WHITE RUSSIAN S.S.R **UC2**

SPAIN .... **EA**
SPANISH SAHARA **EA9**
SRI LANKA **4S7**
SUDAN .... **ST**
SURINAM **P2**
SVALBARD IS .... **JW**
SWAN IS .... **HRO**
SWAZILAND **3D6**
SWEDEN .... **SJ-SM**
SWITZERLAND .... **HB**
SYRIA .... **YK**

TAIWAN .... **BV**
TANZANIA **5H3**
THAILAND .... **HS**
TIBET **AC4**
TIMOR PORTUGUESE .... **YB-YD**
TOGO **5V**
TOKELAU IS **2M7**
TUNGA **A35**
TRANSKEI **S8**
TRANDADE & MARTIM VAZ IS .... **PYO**
TRINIDAD & TOBAGO IS **9Y4**
TRISTAN DE CUNHA **ZD9**
TROMELIN IS **FR7**
TUNISIA **3V8**
TURKEY .... **TA,TC**
TURKS & CAICOS IS **VP5**
TUSCAN ARCHIPELAGO .... **IA**
TUVALU IS **VR8**

UGANDA **5X5**
UNITED ARAB EMIRATES **A6X**
UNITED NATIONS, GENEVA **4U**
UNITED STATES OF AMERICA .... **N,W,K**
URUGUAY .... **CX**

VITICAN CITY .... **HV**
VENEZULA .... **YV**
VIETNAM **XV5**
VIRGIN IS **KV4**
VOLTAIC REPUBLIC **XT2**

WAKE IS **KW6**
WALES .... **GW**
WALLIS & FUTUNA IS **FW8**
WILLIS IS **VK9Z**

WINDWARD IS:
- DOMINICA **VP2D**
- ST. LUCIA **VP2L**
- ST. VINCIENT & DEPENDENCIES **VP2S**

YEMEN **4W**
YUGOSLAVIA .... **YU**

ZAIRE, REPUBLIC OF **9Q5**
SAMBIA **9I,9J**
ZANZIBAR (TANZANIA) **5H1**

Index

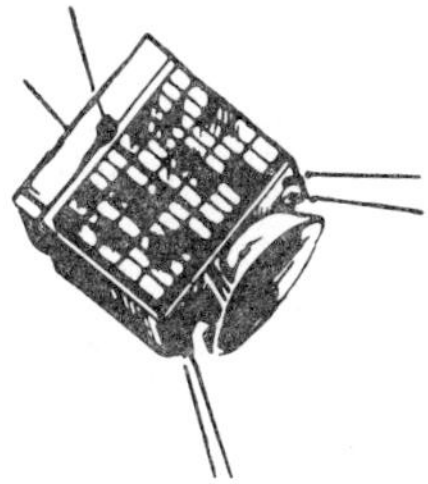